416 880-6243

542 7393

# About Island Press

Since 1984, the nonprofit Island Press has been stimulating, shaping, and communicating the ideas that are essential for solving environmental problems worldwide. With more than 800 titles in print and some 40 new releases each year, we are the nation's leading publisher on environmental issues. We identify innovative thinkers and emerging trends in the environmental field. We work with world-renowned experts and authors to develop cross-disciplinary solutions to environmental challenges.

Island Press designs and implements coordinated book publication campaigns in order to communicate our critical messages in print, in person, and online using the latest technologies, programs, and the media. Our goal: to reach targeted audiences—scientists, policymakers, environmental advocates, the media, and concerned citizens—who can and will take action to protect the plants and animals that enrich our world, the ecosystems we need to survive, the water we drink, and the air we breathe.

Island Press gratefully acknowledges the support of its work by the Agua Fund, Inc., The Margaret A. Cargill Foundation, Betsy and Jesse Fink Foundation, The William and Flora Hewlett Foundation, The Kresge Foundation, The Forrest and Frances Lattner Foundation, The Andrew W. Mellon Foundation, The Curtis and Edith Munson Foundation, The Overbrook Foundation, The David and Lucile Packard Foundation, The Summit Foundation, Trust for Architectural Easements, The Winslow Foundation, and other generous donors.

The opinions expressed in this book are those of the author(s) and do not necessarily reflect the views of our donors.

# Water Resources

**Foundations of Contemporary Environmental Studies**

James Gustave Speth, editor

*Global Environmental Governance*
James Gustave Speth and Peter M. Haas

*Ecology and Ecosystem Conservation*
Oswald J. Schmitz

*Markets and the Environment*
Nathaniel O. Keohane and Sheila M. Olmstead

Forthcoming:

*Coastal Governance*
Richard Burroughs

# WATER RESOURCES

## Shimon C. Anisfeld

WASHINGTON | COVELO | LONDON

ISLAND PRESS is a trademark of the Center for Resource Economics.

Library of Congress Cataloging-in-Publication Data
Anisfeld, Shimon C.
     Water resources / by Shimon C. Anisfeld.
        p.    cm. — (Foundations of contemporary environmental studies series)
     Includes bibliographical references and index.
     ISBN-13: 978-1-59726-494-5 (cloth : alk. paper)
     ISBN-10: 1-59726-494-6 (cloth : alk. paper)
     ISBN-13: 978-1-59726-495-2 (pbk. : alk. paper)
     ISBN-10: 1-59726-495-4 (pbk. : alk. paper) 1. Watershed management. 2. Water resources development—Environmental aspects. 3. Water-supply—Social aspects.
I. Title.
     TC409.A73 2010
     333.91—dc22                                                    2010002475

Printed on recycled, acid-free paper

Manufactured in the United States of America
10   9   8   7   6   5   4   3   2

Keywords: watersheds, floods, scarcity, drought, pollution, sanitation, irrigation, energy, water markets, privatization, water management

*For my parents, who gave me—among other things—
the desire to learn, to teach, and to write*

# Contents

# Preface

This book is a concise, but reasonably comprehensive, introduction to the science and policy of water resources. It is meant to provide everything you need to know for a basic understanding of the global water crisis and how we might move toward solving it. It is targeted at three audiences: students, professionals, and laypeople.

In a university setting, this book can be used as the main text for a graduate-level water resource management course, preferably supplemented with readings from the primary literature; suggested readings are provided for each chapter. It can also be used, together with the other books in this series, in an undergraduate introduction to environmental science and policy.

For environmental professionals without extensive water experience, this book can serve as a way to "get up to speed" on water issues. For people working with water issues every day, this book can provide a larger context and serve as a handy up-to-date reference.

This book can also help interested laypeople understand the basics of water, including how their own actions affect, and are affected by, water resources. Given the pervasiveness of water in all our lives, I strongly believe that every citizen should have some understanding of the issues surrounding this essential resource.

Why is it important to study water? As we will see throughout the book, water is a resource that is both vital and threatened. Our lives and our prosperity are absolutely dependent on having a safe, adequate supply of water for use in our homes, our factories, and our farms. Yet both human water systems and the natural ecosystems that underlie them are under threat from all directions: scarcity, pollution, wasteful use, environmental degradation, global climate change, water conflict, and inadequate human access to safe water and sanitation.

In order to ease this growing water crisis, we must improve the management of our water resources. Thus, in addition to examining the current water crisis and its predicted worsening over the next several decades, this book also looks at some ways that these dismal predictions might be averted—some tools that might help us move toward a better future.

Here are the main elements of the approach that I take in this book:

**Disputes and dialectics.** In the water field, many areas of both science and policy are genuinely uncertain and disputed. A case in point on the science side is the projected impact of climate change on water availability, while a policy example is the role that markets should play in allocating water. Although I often provide my own judgment on the relative merits of different arguments, I believe that it is also important to fully explain both sides and give readers the tools to understand the argument and make their own decisions.

**Data.** When faced with the disputes mentioned above, I have tried, as much as possible, to turn to real data to help decide among different positions. Likewise, I have tried to present as much hard evidence as possible to address various questions throughout the book, ranging from the current status of aquatic ecosystems to the best ways to stop the cycle of disease transmission in developing countries. Of course, that has not always been possible, both because of space constraints and because we lack sufficient data to evaluate many issues. Uncertainty pervades the water field, as it does in any field where science is actively trying to address cutting-edge questions. I have tried to make readers cognizant of uncertainties as we try to piece together a picture based on the available evidence.

**Details.** I believe that understanding the details is important, even in an overview book. Throughout the book, I try to introduce readers to the terminology, assumptions, and approaches used by different disciplines. My goal is to make readers truly literate in this field, able to read the primary literature or attend a conference with confidence in their ability to cut through the jargon and understand the points being made. Some of the issues I discuss are moderately complex and technical, but I have tried to explain them systematically and carefully, so that even readers with no background in the field can work their way through the material.

**Disciplines.** Water is inherently an interdisciplinary subject. It draws on hydrology, chemistry, ecology, geomorphology, climate science, economics, law, sociology, and the policy sciences. My own background is in the natural sciences, but I have tried—with the generous help of colleagues— to bring multiple perspectives to bear in this book. Of course, it is impossible in a book of this scope (or probably any single book) to genuinely represent the framework and contributions of each of these fields, and I refer readers to a variety of disciplinary textbooks for further exploration (see Recommended Readings).

**Linkages.** Is this a book about environmental protection or a book about human resource use? It is both, of course, but the question points to one of the central tensions that underlies this book: the back-and-forth between concern for human society and concern for ecosystems. I strongly believe that the two are linked: that ecosystem degradation ultimately affects the viability of our economies and our lives, and that economic development must respect the environment if it is to be successful in the long term (the *sustainable development* perspective). This human/environment linkage is a theme that runs throughout the book and is most fully developed in chapter 8.

I would like to express my deep gratitude to the people who helped me in the writing of this book: Emily Davis and Todd Baldwin, my editors at Island Press, for their guidance in shaping the book and for their trust in me (despite many missed deadlines!); Gus Speth for providing me with the opportunity to write this book and for encouraging me to think deeply about water; Sheila Olmstead for providing an economist's perspective and making significant improvements to Chapter 12; Brad Gentry for providing useful feedback on Chapter 12 and helping guide me through the complexities of international financing; Mark Ashton for patiently correcting my descriptions of forest issues in sections 6.2 and 11.2; Chris Bellucci and Paul Stacey of the Connecticut Department of Environmental Protection for helpful comments on Box 8.3; Edouard Pérard of the World Bank for sharing the data presented in Figure 12.1; Azalea Mitch of the Greater New Haven Water Pollution Control Authority for her thoughts on Clean Water Act enforcement; Jeff Albert of the Aquaya Institute for his thoughts on water conflict; and all my students and colleagues at Yale and elsewhere for teaching me so much about water resources.

Stacey Maples, Yale's master mapmaker, produced 10 wonderful maps that greatly enriched the book; I thank him for all his hard work and for his patience with my changes.

Special thanks go to Elizabeth Anisfeld for reading the entire manuscript and providing helpful, detailed edits along with constant encouragement, and to Sharon Cohen Anisfeld and Moshe Anisfeld for commenting thoughtfully on several chapters.

Lastly, I thank Sharon, Daniel, and Tali for encouraging and supporting me, for putting up with my foul moods when the writing was not going well, and most of all for their ongoing love and companionship.

# 1

## Past, Present, and Future: Introduction to the Water Crisis

Take a moment to imagine water, to bring a picture of this vital substance to your mind's eye.

What do you think of? A favorite lake or river? A glass of drinking water? A refreshing shower? An irrigation sprinkler? A baptismal font?

People and water are deeply interlinked. We humans are mostly water, as is the surface of the planet we live on. We use water in so many ways: drinking; cleaning ourselves and our belongings; growing our food; making the products we use; producing energy; enjoying ourselves while swimming or boating; fishing; transporting people and goods; sustaining our spirits. Equally important to our survival, but less obvious, is the fact that water plays a critical role in maintaining the health of the earth by controlling the planet's heat and energy balances, cycling nutrients and other elements, allowing the growth of the plants and algae on which all life depends, and maintaining the biodiversity of aquatic and terrestrial ecosystems.

This book is an introduction to water resources, in which we explore the many interactions between people and water. It emphasizes two contrasting—but complementary—ideas: (1) we are facing serious water problems, and (2) there is much that we can do to manage this resource better and alleviate those problems. This chapter provides a brief history of past water management, a summary of the present crisis, and some thoughts on possible paths toward a better future.

## 1.1. The Past: A Brief History of Water Resource Management

Humans have always needed water, and early civilizations developed where water sources were available to support their populations. Over time, different regions developed their own approaches to managing water, displaying a rich diversity of technologies. A few examples: The ancient Egyptians were heavily dependent on the Nile for irrigation water, which they obtained both through natural flooding and through active water management such as the digging of *irrigation canals* and the utilization of water-lifting devices such as the *shadouf.* The Egyptians also developed a system for monitoring and recording river water levels. In Persia, tens of thousands of gently sloping underground tunnels called *qanats* were dug to deliver groundwater to cities and fields; similar technologies with different names have been used around the world (*karez* in Afghanistan, *laoumia* in Cyprus, *surangam* in India). The Romans built *pipelines* and large raised *aqueducts* to transport water, and constructed buried *sewer pipes* to remove wastewater. The Chinese constructed irrigation canals to bring water from rivers to fields, dug the 1600-km-long Grand Canal to transport goods and people, and built *levees* along the Huang He (Yellow River) to protect land from flooding. The Spanish developed a system of irrigation canals called *acequias*, which they later introduced to the areas that they colonized in the New World. In England and elsewhere, water was used to power *mills*, often by impounding water behind a small *dam.*

Along with these technologies came cultures and governance systems for managing water. The historian Karl Wittfogel famously argued that water had a strong influence on the development of despotic civilizations ("hydraulic empires") in the Middle East and Asia, due to the need to harness large amounts of manpower to build and maintain water delivery systems in arid regions with large rivers (Wittfogel 1957). Wittfogel's thesis is now considered overly simplistic, but it points to the ways that water—as arguably the most important natural resource of all—can influence the development of civilizations.

### The Twentieth Century

The twentieth century saw global population increase dramatically from about 1.6 billion to 6.1 billion. This population explosion, together with the development of new technologies, led to significant changes in water management. In turn, these changes in water management were crucial in allowing the expansion in population, which would not have been possible without new water supplies for cities and for agriculture.

Water management in the twentieth century was dominated by what has been referred to as the "hard path." Three technologies played crucial roles in defining this path.[1]

## Dams

For centuries, small dams have powered mills and impounded water for storage and delivery. But the twentieth century saw an explosion in dam-building, especially of *large dams*—those over 15 meters high—and *major dams*—those over 150 meters high.[2] The real era of dam-building began in the mid-twentieth century, first in the United States and then in other countries. By the year 2000, there were approximately 48,000 large dams worldwide, an astonishing number. Although North America and Europe have largely stopped building dams, dam construction in Asia continues apace, especially in China and India. China now has about half of all the large dams in the world (WCD 2000).

Dams are built for several purposes:

- Water supply and irrigation: Dams allow for the delivery of water to urban ("water supply") and agricultural users, through two mechanisms. First, dams simplify the process of capturing and delivering water from a river by creating an impoundment from which to draw water and by increasing the elevation of the water so that it can be delivered more easily by gravity. Second, dams capture water during wet periods and store it for delivery during dry periods. This is particularly important in seasonal or drought-prone climates.
- Flood control: Dams can capture large floods and release them slowly to minimize flooding damage downstream.
- Hydropower: The power of falling water can be used to run watermills or, as is done today, drive turbines that generate electricity.
- Navigation: Dams convert flowing, shallow water to still, deep water, which can allow ships and barges to more easily move up and down river systems.

---

1. It could be argued that levees and water treatment plants are equally important components of the hard path.

2. Technically, a dam is classified large if it meets one of the following criteria: height >15m or reservoir volume >3 million cubic meters (MCM). A dam is considered major if it meets one of these criteria: height >150m, reservoir volume >25,000 MCM, dam volume >15 MCM, or installed hydroelectric capacity >1000 megawatts.

- Recreation: A secondary purpose for many large dams is water-based recreation on the reservoir that develops behind the dam.
- Fishing: Usually a secondary purpose, some reservoirs are used for capture fisheries or aquaculture.

Many of the largest dams serve multiple purposes, though managing for these different purposes may require divergent approaches. For example, managing for flood control involves keeping the reservoir as low as possible between storm events, while managing for water supply involves storing as much water as possible.

Beyond these specific purposes, large dams are symbols of development, technology, and "the control of nature." The Hoover Dam, for example, symbolized technological optimism to an America struggling to emerge from the Great Depression. The awe inspired by the Hoover Dam is captured in President Franklin Delano Roosevelt's response to seeing the immensity of the structure: "I came, I saw, I was conquered." Likewise, the ability to successfully pull off a large dam project has been a rite of passage and point of pride for many developing nations.

To many people, however, dams symbolize exactly what is wrong with the hard path: the brutal attempt to control nature with raw force; the lack of respect for free-flowing rivers and healthy ecosystems; the disregard for the lives of local citizens displaced by the reservoir and dam workers injured or killed during construction; the disease of "gigantism" that tries to build its way out of every problem.

In the US, two federal agencies have been particularly responsible for building and operating large dams: the Bureau of Reclamation, established by the Reclamation Act of 1902 and given the mission of helping settle the arid West by supplying irrigation water to farmers (and, later, hydroelectric power to cities); and the Army Corps of Engineers, which has responsibility for flood control, navigation, and sometimes water supply in much of the eastern half of the country.

## Canals

Two types of artificial waterways have played key roles in the hard path of water management: conveyance canals and navigation canals.

If dams gave us the ability to store water, *conveyance canals* (along with pipes and tunnels) gave us the ability to deliver it over large distances to cities and farms that have sprung up far from any natural sources of water. Examples that stand out include the Central Arizona Project (540 km), which delivers water from the Colorado River to Phoenix and Tucson and to farmers in central Arizona; the All-American Canal (130 km), which transports water from the Colorado to the Imperial Irrigation District in California; and the plan under way in China to move massive amounts of water from south to north. In many cases, conveyance canals are used to move water from one river basin to another (*interbasin water transfer*), which can lead to high environmental, economic, and social costs. One source (Thatte 2007) estimates that there are currently 134 interbasin transfers worldwide, representing about 14% of the world's water use; proposed projects (including the Chinese project mentioned above) would increase the volume of water moved by a factor of 3.[3]

*Navigation canals* allow the inexpensive movement of people and goods over large distances. Whereas navigation on rivers involves adjusting our mobility to the vagaries of the natural river network, the construction of navigation canals allows us to adjust the water network to our transportation needs.

## Wells

The third member of the hard path's technology trio—the well—is perhaps the least obvious, because it is the smallest and most widely distributed. Hand-dug wells have been used for millennia as a way to obtain water, but our use of wells and *boreholes* (narrow drilled wells) grew tremendously during the twentieth century, due to advances in drilling technology and the availability of electric and diesel-powered pumps. Untold millions of wells are now used to extract water at high rates from great depths. This has allowed us to tap into groundwater in a way that was previously impossible. In many places, we are now using groundwater much more quickly than it is being replenished—meeting today's water needs at the expense of future generations.

## 1.2. The Present: A Water Crisis

Despite the tremendous technological advances of the twentieth century, most water experts agree that we are now facing an unprecedented global water crisis. This crisis can be broken into 10 interrelated components, each of which will be covered in a chapter of this book.

Flooding (Ch. 4). Despite our efforts to protect ourselves from flooding, large floods still do extensive—even increasing—damage to property and people.

---

3. These estimates do not include the many thousands of interbasin transfers that are currently operating at smaller scales.

Scarcity (Ch. 5). More and more regions are starting to run out of water as population grows, per-capita water use increases, and pollution renders water sources unusable.

Change (Ch. 6). Several interacting dynamics—population growth, climate change, and land use change—are altering both the supply of water and the demand for it, and are posing serious challenges to water management methods that are based on the patterns of the past.

Technology (Ch. 7). While appropriate technologies are an important part of the solution, the indiscriminate construction of large water infrastructure projects is part of the problem, causing serious environmental and social impacts.

Ecosystem degradation (Ch. 8). Aquatic ecosystems have paid a heavy price for our water management, and we are in danger of losing the natural foundation that underlies our physical, economic, and spiritual well-being.

Human health (Ch. 9). Billions of people in developing countries lack access to safe water and adequate sanitation, and suffer from serious health consequences as a result; even in rich countries, drinking water can contain toxics and pathogens.

Agriculture (Ch. 10). Our ability to grow enough food for the whole world is threatened by lack of water; at the same time, agricultural land is being degraded by salinization and pollution.

Industry (Ch. 11). Energy use and industrial activity require large volumes of water and contribute to the pollution and degradation of water resources.

Inefficiency and inequity (Ch. 12). Current water policies often allocate water inefficiently and inequitably, with some users being granted excessive volumes of cheap water, while others lack sufficient water for vital needs.

Conflict (Ch. 13). Throughout the world, scarce water is leading to tensions among different sectors, states, and countries; these tensions, in turn, get in the way of more efficient, cooperative water management.

## 1.3. The Future: An Emerging Approach to Water Management

We are now at a crucial juncture in water management. The challenges of the twenty-first century demand a new approach that builds on the lessons of the past and incorporates innovative philosophies and technologies. This emerging approach to water management is finding expression in three related philosophies: the "soft path," Integrated Water Resource Management, and the Watershed Approach.

The soft path approach—in contrast to the "hard path" of building large, centralized infrastructure to meet ever-growing demands for water—focuses on water management that accommodates both human and ecosystem needs through a combination of conservation, decentralized technologies, and integrated management. The term *soft path* was originally used in the context of energy choices by Amory Lovins (1977) and was first used in the water context by Brooks (1993). However, it was popularized and more fully developed for water by Peter Gleick (Gleick 2002, Wolff and Gleick 2002).

Integrated Water Resource Management (IWRM) originates from a *sustainable development* context, which emphasizes meeting current development needs without impairing the integrity of natural systems and their ability to meet future needs. The term IWRM was first articulated internationally in the 1992 Dublin Statement on Water and Sustainable Development, which came out of a preparatory meeting for the United Nations Conference on Environment and Development (UNCED) in Rio de Janeiro. IWRM is currently a key component of international water discourse, although it seems to mean somewhat different things to different people.

The Watershed Approach is essentially the US Environmental Protection Agency's version of IWRM. The Watershed Approach was initiated in 1996, with further guidance published in 2001 and 2008. States throughout the US are currently creating and implementing watershed plans based on this approach.

Below I outline the key features of the emerging approach to water management, drawing on what I believe to be the most important contributions of each of the philosophies mentioned above.

Balancing Different Demands. As noted in section 1.1, humans rely on water resources for a variety of different uses—uses that often exert conflicting demands on these resources. In the past, water development has often proceeded in a single-sector fashion, with each group of users implementing their own plans without coordination with others, resulting in both conflict and inefficiency. The key challenge of water management is figuring out how to best balance all the different demands on a water resource, from drinking water to navigation to ecosystem protection. Integrated planning that considers all users is very difficult to do but can lead to much better outcomes.

Efficiency and Equity. The values to be considered in balancing different demands can be summed up by the "two Es": efficiency—obtaining the maximum total benefits from the water resource; and equity—distributing those benefits in a fair way.

Sustainability. In evaluating the demands on a water resource, it is important to include two key "users" that are often ignored:

- Future users: The principles of sustainable development demand that today's water uses not compromise the needs of future generations.
- Ecosystems: Although it would be unrealistic to prohibit any harm to ecosystems, ways must be found to minimize this harm and to use water in ways that maintain ecosystem integrity to the greatest possible extent. This is sometimes expressed by adding a third "E"—environment—to the two Es above. In addition, two important terms have been used to capture essential concepts related to environmental sustainability. *Ecosystem services* refers to the fact that healthy ecosystems provide critical services to human society, such as clean water and fisheries. *Natural infrastructure* conveys the sense that intact ecosystems can substitute for investment in infrastructure; for example, watershed protection can alleviate the need for building a water treatment plant.

Including Stakeholders in Decision Making. A corollary to the requirement to consider all users is the need to include all users early in the planning and decision-making process. The stakeholder groups that need to be identified and engaged include groups representing particular users (e.g., irrigation districts, environmental groups); government agencies that are tasked with regulating or promoting the interests of certain users (e.g., Department of Agriculture, Environmental Protection Agency, state and local governments); those with expertise that can help move the process along (natural and social scientists, engineers); and those who will shape the public perception of the process (media). It is also important to include two groups that may not view themselves as water users, but can have a significant impact on the resource, namely, polluters who discharge to surface or groundwater, and landowners whose activities may affect the quantity and quality of runoff.

Utilizing Local Wisdom. In the past, national governments or international groups have often conducted water resource planning in isolation from local populations and interest groups. When it came time to implement these plans, they often failed because they did not anticipate the physical or social realities on the ground. Methods and technologies that are adapted to the particular constraints and opportunities of a given watershed or community are more likely to succeed than one-size-fits-all approaches. Future water planning and implementation must incorporate both local wisdom and international expertise, both the expectations of the community and the constraints of national and international law.

Attention to Governance and Management. In many countries, the era of "infrastructure investment" is ending and the era of "management investment" is beginning (Briscoe and Malik 2006). That is, we have built enough infrastructure that the return on new construction will be small relative to the return that we will get from better management of existing infrastructure, both constructed and natural. The challenge now is to operate water systems more efficiently, by focusing on maintenance, professional training, sound operational procedures, attention to user needs, and healthy governance structures.

Providing Water Services, Rather than Water. A key insight of the new approach is that people don't necessarily want to use water per se; rather, they want the services that water provides. For example, farmers don't want to use water, they want to grow crops; if we can provide them with ways to achieve the same yields with less water, they will be happy to do so. Likewise, when we flush the toilet, we are not interested in using water but in safely disposing of human waste; if alternative technologies can do so with reduced water use (or no water at all), that may make little difference to the end user—although cost, convenience, and cultural acceptance of the new technology must certainly be considered.

Thinking About Water Quality. Not all water is the same, and considering water quality as well as quantity is critical in water resource planning. One way to take advantage of this factor is to match waters of different qualities to different end uses. For example, do we really need to use potable water to flush toilets, or could we use lower-quality water for this purpose, if a supply of such water were available?

Identifying the Right Geographic Scale. Water management has a strong geographic focus: it involves balancing all the competing demands within the context of a particular geographic area and its water resources. The choice of geographic unit can be complicated, but should be physically and socially coherent, and should encompass both surface and ground-water when relevant. The watershed or river basin is often a good choice, since it delineates a well-defined shared water resource. Yet it is not always the right scale, for reasons we will discuss in Chapter 2.

These key principles will play out in different contexts throughout the book.

## 1.4. Structure of This Book

Besides this introductory chapter and a very short concluding chapter, this book consists of 12 chapters. The first two (Chs. 2 and 3) provide basic material necessary for understanding the remainder of the book: an introduction to hydrology and water quality (Ch. 2) and an overview of global water resources and water use (Ch. 3). We then turn to 10 chapters corresponding to the components of the water crisis discussed above. Although each chapter deals with a different topic, they should be read in order, as they build on each other in a logical sequence.

Readers are encouraged to avail themselves of several resources located in the back of the book: references for following up on specific material cited in the text; recommended readings for each chapter and for the water field as a whole; discussion questions meant to provoke further reflection on the material covered in each chapter; and a glossary for quick reference, a feature that may be quite helpful given the numerous acronyms and technical terms used throughout the book.

# 2

# Quantity and Quality: Introduction to Water Resource Science

In order to understand water resource issues, we need to first have a basic knowledge of the physical principles governing water movement in the environment and of the chemical principles underlying water quality.

This chapter introduces the basics of hydrology and water quality, important scientific background that we will use throughout the book. We also touch very briefly on the most relevant principles of two other scientific disciplines, fluvial geomorphology and aquatic ecology.

We begin the chapter with an overview of the hydrologic cycle, then look closely at the watershed concept and how it can be applied at different scales, provide a primer on water quality, and finally define and discuss geomorphology and ecology.

## 2.1. Following the Hydrologic Cycle

We begin our exploration of water science by investigating the movement of water in the environment, known as the *hydrologic cycle*. The term *cycle* reflects the fact that the amount of water on Earth is fixed (with the exception of small inputs of water from space), but this water is continually moving from ocean to atmosphere to soil to river and back again. In this section, we follow a drop of water on its course through this cycle.

### Precipitation

We could start our discussion anywhere in the cycle, but one natural place to begin is with a drop of water falling from the sky as *precipitation*.

Precipitation (e.g., rain, sleet, hail, snow) generally occurs as a result of *adiabatic cooling*, a process in which an air mass rises, expands, and cools. Since cool air can hold less water than warm air, the cooled air mass can become *supersaturated* with water vapor, leading to formation of clouds and drops of condensed water which then fall to the ground.

Different types of air mass movement can lead to precipitation with different characteristics:

- Orographic: Where moisture-laden air encounters mountains, it is forced upward, leading to adiabatic cooling and precipitation on the windward side of the mountains. On the lee of the mountains, there will typically be little rain, since the descending air mass (now undergoing adiabatic heating) can hold much more moisture than it has left.
- Frontal: Precipitation often forms along the boundaries of air masses of different temperatures and densities. A cold front moving under warmer air will lead to relatively short, high-intensity rainfall, whereas a warm front displacing cooler air will lead to more extended, lower-intensity precipitation. Tropical cyclones (also known as hurricanes) are low-pressure systems that can drop massive amounts of rain in a short time.
- Convective: On warm afternoons, heating of air near the surface can cause it to rise and lead to short, intense precipitation, often in the form of thunderstorms.

The amount of precipitation can be measured by *rain gauges* or can be estimated by radar. A graph of precipitation over time is referred to as a *hyetograph*. Precipitation is generally expressed in depth per unit time (e.g., mm/day) or total depth over a given event (e.g., mm).

As our drop of water falls, it may either hit the ground directly (*throughfall*) or encounter a vegetated surface, such as a tree leaf (*interception*). The intercepted moisture on the vegetation will eventually either flow down the plant to the ground surface (*stemflow*) or, more likely, evaporate from the plant without ever reaching the ground.

Upon reaching the ground, our water drop may either run off the surface or *infiltrate* into the soil and become part of the *soil moisture* pool. The maximum amount of water that the soil can hold is determined by the amount of void space in the soil, known as the *porosity* (a dimensionless ratio: pore volume divided by total volume). When the pore space is completely occupied with water (rather than air), the soil is said to be *saturated*. More typically, however, the soil near the surface is *unsaturated*, with a *volumetric moisture content* (volume of water divided by total volume) somewhere between zero and the porosity.

## Evapotranspiration

Our drop of water can return to the atmosphere from soil or a water body either by *evaporation* (the physical movement of water from liquid state to gas state[1]) or by *transpiration* (the biological movement of water from the soil into the plant and then out from the leaves as water vapor). Transpiration is essential for plant growth, because in order to take up $CO_2$ for photosynthesis, plants must open their stomates and in the process release water vapor. Evaporation and transpiration are often referred to collectively as *evapotranspiration*, or ET, generally expressed in depth per unit time (e.g., mm/day).

ET is extremely hard to measure over any sizable area. It is often estimated using a hypothetical construct known as *potential evapotranspiration*, or PET. PET is meant to reflect the potential for water vaporization in a given time and place—if water were abundant. It can be approximated by *pan evaporation*, that is, the loss of water from a water-filled pan of defined size and shape. PET is controlled by temperature, humidity, wind, and solar radiation, with the highest PET coming under hot, dry, windy, and sunny conditions. In the real world, ET will generally be less than PET, since low water availability in the soil will lead to lower evaporation than the theoretical potential. In addition, the presence and type of vegetation will affect ET (by controlling the rate of transpiration) but not PET.

## Percolation and Groundwater

Water that does not evaporate from the soil may percolate downward through the unsaturated zone (also referred to as the *vadose zone*) and reach the *groundwater zone*, defined simply as the area where all the pores are saturated with water—or, in the case of bedrock, where any fractures in the rock are saturated with water.[2] This percolation of water from above is referred to as *diffuse recharge* of groundwater. The *water table*—defined as the top of the groundwater zone—may be near the surface or may be at great depth. Once it is part of groundwater, our drop of water

---

1. When water moves directly from solid (ice or snow) to gas, the process is referred to as *sublimation*.

2. More precisely, there is a zone of saturation above the groundwater zone, referred to as the *capillary fringe*, where water is pulled upward by capillary forces having to do with the surface tension of water in small pores.

will move laterally in response to differences in *hydraulic head* (essentially differences in the elevation of the water table).

, The two most important characteristics of a groundwater-containing formation are the amount of water that it contains, which is described by the porosity; and the ease with which that water will move, which is described by a parameter called the *hydraulic conductivity*. Generally, larger pores—such as those found where the subsurface is composed of sand and gravel—have higher conductivities (water moves more readily) and make up formations referred to as *aquifers*, from which it is relatively easy to pump out water. In contrast, smaller pores—such as those found in silt, clay, or mixed-material formations—hold on to water more tightly and form *aquitards* (low conductivity) or *aquicludes* (lowest conductivity). Aquifers that are overlain by the vadose zone (as described above) are referred to as *unconfined aquifers* and are under atmospheric pressure. In contrast, *confined aquifers* are overlain by an aquiclude, which prevents water flow through it; the water in a confined aquifer is often under high pressure.

Groundwater and surface water can interact in a variety of ways. When groundwater reaches the surface at a stream, lake, or wetland, the direction of flow may be either from the groundwater to the surface water (*discharge* of groundwater to surface water) or from the surface water to the groundwater (*localized recharge* of groundwater from surface water). In some regions (especially humid areas), groundwater and surface water are closely linked, with groundwater serving primarily as a storage location for water en route from soil to stream—or, to put it differently, diffuse recharge of groundwater from precipitation is balanced by discharge of groundwater to streams. In other locations, the water table is much deeper, and groundwater never reaches the surface to discharge to a stream. In many of these locations, recharge is also very low and the groundwater represents a "fossil" or nonrenewable resource that was formed over many years and/or under different climatic conditions.

## Streamflow

After entering groundwater, our drop of water may move laterally and ultimately enter a stream, as noted above. *Streamflow*, or *discharge*, is the volume of water that a stream carries past a given location per unit time (units: $m^3/sec$ or $ft^3/sec$, also known as cfs [*cubic feet per second*]). A graph of flow over time is referred to as a *hydrograph*; the individual time points that make up the hydrograph typically represent average flows over periods ranging from 15 minutes to a year (Figure 2.1).

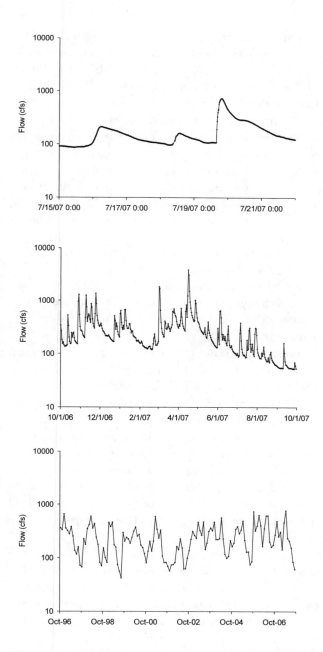

*Figure 2.1.* Hydrographs from the Quinnipiac River (USGS station 01196500) over three different time scales. Top: A week's worth of 15-minute data. Middle: A year's worth of daily data. Bottom: A decade's worth of monthly data. Data source: USGS 2009.

Hydrographs typically show variation at several temporal scales. At the scale of the individual storm event, the flow generally rises in response to rainfall ("the rising limb") and then slowly declines ("the falling limb") over a time period that can range from minutes to weeks, depending on the size of the river and the duration of the rain event. At longer time scales, flow varies seasonally and interannually in response to changes in precipitation, evapotranspiration, and snowmelt. For the river shown in Figure 2.1, precipitation is (on average) evenly distributed through the year, but flows are generally somewhat lower in summer because of higher ET. In contrast, rivers in climates with highly seasonal precipitation (e.g., Ganges River) can show a much larger seasonal range in streamflow. For still other rivers, a spring flow peak due to snowmelt is the dominant hydrologic event.

The flow of water in a stream in direct response to a storm event is referred to as *stormflow*, while flow between storm events is referred to as *baseflow*. The pathways by which water actually flows to the stream can be quite complex, but pathways of stormflow tend to be both more rapid and closer to the surface than baseflow, which tends to be fed by slow discharge of groundwater.

During storms, *overland flow* (water movement along the surface) can be generated in two ways: *infiltration-excess (Hortonian) overland flow*, in which the rain is falling faster than the soil can absorb it; and *saturation-excess overland flow*, in which the rain is falling on saturated soil. The former is typical of very high intensity precipitation events or degraded soils with low infiltration capacity, while the latter is typical of longer events on healthier soils, especially in low-lying areas. Overland flow can also be generated in the immediate vicinity of stream channels by the discharge of groundwater due to a storm-induced rise in the water table (known as *return flow*). *Shallow subsurface stormflow* is usually slower than overland flow, but can sometimes occur surprisingly rapidly in *macropores* that are oriented parallel to the surface.

Where continuous records of streamflow are available (e.g., from the US Geological Survey [USGS]), these are generally obtained using dataloggers that automatically record water depth, which is then converted to streamflow using a *rating curve*. The rating curve is the statistical relationship between water depth and streamflow, and is obtained from periodic measurements of depth and flow under different conditions.

Most rivers ultimately discharge their water into the oceans. However, some rivers discharge into inland areas from which the water evaporates and never reaches the oceans. Examples of these *endorheic regions* include the watersheds of the Great Salt Lake and the Aral Sea.

## 2.2. Watersheds and Issues of Spatial Scale

We now turn to discussing the *watershed* as a basic hydrologic concept and as a useful unit for water resource science and management. As we will see, understanding the watershed concept is closely tied to understanding issues of spatial scale in water resources.

### Definition

A watershed can be defined as "the topographic area within which apparent surface water runoff drains to a specific point on a stream or to a waterbody such as a lake" (Omernik and Bailey 1997). In other words, the watershed that is defined by a given point is simply the land area that contributes water to that point. Watersheds are delineated based on surface topography, with the understanding that water that falls as precipitation will flow downhill to the stream and then flow downstream past our selected point. Of course, not all the precipitation that falls in the watershed will necessarily become streamflow (some will evaporate), but all of the precipitation that does become streamflow must pass by our point. Delineating a watershed, then, is simply an exercise in tracing gravity, although this exercise can be quite difficult (or impossible) in areas that are extremely flat. In common parlance, the term watershed (as in "a watershed moment") is also used to refer to what hydrologists would call a *watershed divide*, the high ridge that separates two watersheds from each other.

How does groundwater play into the definition of a watershed? The watershed concept is most useful in situations where groundwater is closely associated with surface water, that is, where groundwater simply serves as storage for water en route from soil to stream. In such cases, we can think of groundwater as following the same boundaries as the surface watershed and can treat it as part of the watershed. In these situations, water withdrawn from groundwater wells will reduce surface water flow within the watershed and should be treated by managers similarly to surface water withdrawals. In other situations, however, as discussed above, groundwater forms a separate resource that is not closely linked to surface water. Where that is the case, groundwater may follow very different boundaries from the surface watersheds and should be delineated separately based on aquifer boundaries and subsurface flow patterns.

An important exception to the simple definition of watersheds occurs in areas where humans have significantly influenced water movement so that the surface topography is not the best guide to water flow. For example, large cities commonly transport water across watershed boundaries. What watershed is the city of Boston in? Is it restricted to its relatively small topographic watershed, or does it include the watershed of the Quabbin Reservoir in western Massachusetts, whose waters are piped to Boston residences and end up flowing into Boston Harbor?

## Watersheds and Water Budgets

Why is the watershed concept useful? Because it links the stream back to the land that influences it. And because it is the basic unit within which an important part of the hydrologic cycle takes place. These links become clearest in the construction of *water budgets*.

A water budget is an exercise in tracking the inputs and outputs of water to a defined area (the *budget area*). Given that water cannot be created or destroyed (except through chemical reactions, which are insignificant at our scale), any difference between inputs and outputs must be reflected in a change in the storage of water within the budget area. In general, a water budget looks something like this:

$$\text{inputs} = \text{outputs} + \text{change in storage}$$
$$P + S_i + G_i = S_o + G_o + ET + \Delta S \qquad \text{(equation 1)}$$

where:

$P$ = precipitation
$S_i$ = inflow of surface water to our budget area
$G_i$ = inflow of groundwater to our budget area
$S_o$ = outflow of surface water from our budget area
$G_o$ = outflow of groundwater from our budget area
$ET$ = evapotranspiration of water from our budget area
$\Delta S$ = change in storage of water within our budget area (e.g., change in soil moisture)

The budget is done over a defined time period (often ranging from a day to a year or more).

A word on units: Each of the terms in the budget is expressed in units of volume (e.g., $m^3$). Alternatively, this volume can be divided by the area involved to yield a depth of water (e.g., $m^3$ of water $\div$ $m^2$ of budget area = m of water). Sometimes the time period is included in the units so that the terms in the budget are expressed as volume per unit time (e.g.,

$m^3/yr$) or depth per time (m/yr). The latter (depth per time, m/yr) is equivalent to flow per unit area ($m^3$/sec of flow ÷ $m^2$ of budget area = m/yr, after unit conversions); the standard English units for this purpose are, somewhat confusingly, known as csm (cubic feet per second per square mile of contributing area, $ft^3 \ sec^{-1} \ mi^{-2}$).

If we choose our budget area to be a watershed, we can greatly simplify the budgeting exercise. First of all, by definition, there are no inputs of surface water to a watershed. In addition, in a situation where groundwater and surface water are closely linked, there will be no inputs of groundwater—and also no outputs of groundwater, since all groundwater discharge will be to the stream.[3] Outputs of surface water from the watershed occur at only one location: the point on the stream that defines the watershed (known as the *watershed outlet*). If we choose our budget time period wisely (e.g., a year), we can also make the simplifying assumption that the amount of water stored in the watershed will not change significantly over that time period. In other words, we assume that soil moisture, water tables, and lake levels are all roughly the same as they were on the same date a year ago. In that case, the equation above simplifies to

$$P = R + ET \qquad \text{(equation 2)}$$

where R is runoff (outflow of surface water at the watershed outlet, equivalent to $S_o$ in equation 1). In other words, precipitation must either undergo ET or appear as streamflow at the watershed outlet.

The difference between equations 1 and 2 is the difference between doing a water budget for an arbitrarily defined land area (e.g., a state) and doing a water budget for a watershed. The former is difficult, since measuring all the different inflows and outflows is quite time-consuming. The latter is much more manageable, since it involves only measuring precipitation over the watershed and streamflow at the watershed outlet. (ET is either calculated by difference or estimated from PET.)

The advantage of the watershed concept, then, is that it provides a coherent, simplifying unit for analyzing and understanding the availability of water.

There are three additional benefits to the watershed as a unit of management. First, water users within a watershed are, in the most fundamental way, sharing the same well-defined resource. Water that is used

---

3. In situations where groundwater is deep and disconnected from surface water, we can also ignore $G_i$ and $G_o$ by defining our budget area as including the surface and shallow subsurface (vadose zone) but not the groundwater zone.

upstream is not available for use downstream. Pollution that is discharged upstream affects the quality of water downstream.

Second, the watershed is the natural unit within which anthropogenic transport of water has the lowest ecological and economic costs. When water is withdrawn from a stream and used within the same watershed, the return flow from that use (i.e., the portion of the water that is not consumed) remains within the watershed and ultimately flows back to the stream where it came from, thus reducing the ecological impacts of that withdrawal. Financially, it is often cheaper to move water around within a watershed, since an interbasin transfer will require pumping water over the watershed divide (or tunneling through it).

Third, the watershed concept emphasizes the role that the *land* plays in moving water from precipitation to the stream. The watershed is the land area whose surface and subsurface characteristics will affect the quantity and quality of water, whether through determining the amount of ET or controlling the infiltration rate or contributing pollutants. As a result, the group of people involved in affecting and sharing the water resource includes not just water users, but also land users within the watershed.

## Nested Watersheds and Stream Size

If you read closely the definition of watershed, you will note that each point on a stream defines a distinct watershed. That is, there are an infinite number of overlapping watersheds defined by different points. In practice, we tend to delineate watersheds only at certain points on streams, usually either road crossings or natural outlets (where the stream discharges to a larger stream, lake, or ocean). Still, it should be clear that one piece of land may be a part of many different nested watersheds (Figure 2.2) of vastly different sizes, ranging from the watershed of a small stream to the watershed of the Mississippi River or even of the Atlantic Ocean.

Given the range in watershed scales—and the corresponding range in the scales of water resource problems—we need some language for talking explicitly about the sizes of streams and watersheds. The size of a stream can be described in different ways: its width, its depth, some measure of its flow (perhaps average annual flow). However, these measures have one serious disadvantage: they must be actually measured in the field. In contrast, the most common way of describing stream size, namely, *stream order*, can be assessed from a simple topographic map. The rules for determining Strahler stream order are simple (Figure 2.2):

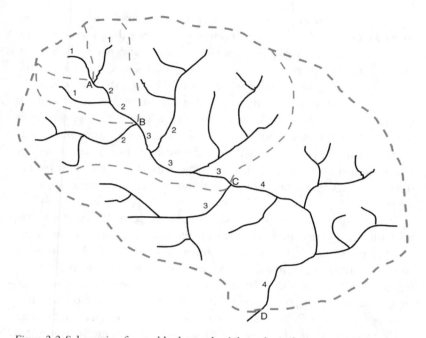

*Figure 2.2.* Schematic of nested hydrographs. A hypothetical stream network is shown, with four levels of nested watersheds delineated. The small watershed defined by point A is nested within the larger watershed defined by point B, which is nested within the larger watershed defined by point C, which is nested within the larger watershed defined by point D. Strahler stream order designations (numbers 1–4) are shown for part of the stream network.

- The smallest perennial stream that is shown on the map is considered *first order.*[4]
- Where two streams of equal order come together, the combined stream has an order that is 1 greater than each of its tributary streams.
- Where two streams of unequal order come together, the combined stream has an order equal to that of the higher-order tributary.

---

4. This means that stream order is dependent on the scale of the map being used. Thus, for example, the Mississippi River at its mouth is considered to be tenth order using USGS topographic maps, but only fifth order using a large-scale global simulated topographic network.

There are many important differences between small streams and large rivers. Hydrologically, the most important difference (besides the obvious fact that large rivers carry more water) is that small streams react much more quickly to storm events, with water levels rising and falling in a relatively short time period. Related to this is the fact that, when expressed in area-normalized units (e.g., mm/sec or csm), peak flood flows in small streams tend to be higher than in large rivers, because the entire watershed contributes runoff at the same time. In a large river, by contrast, the runoff contributions from different parts of the watershed are spread out over time; in addition, it is very unlikely that the most intense storm events (e.g., thunderstorms) will impact the entire watershed of a large river.

Which are more common: small or large streams? As you might expect, there are many more small streams than large, both in terms of the number of streams and in terms of stream mileage (Table 2.1). In other words, when walking across the landscape, you are much more likely to encounter a first- or second-order stream than a large river. At the same time, large rivers obviously have *more* in them: more water, but also more fish, more opportunities for navigation or for deriving hydropower, and so on.

Corresponding to the range in stream sizes is a range in watershed sizes. While two streams of the same order can have somewhat different watershed sizes (due to differences in topography, precipitation, stream network topology, etc.), we can calculate average watershed sizes for each order of stream, as shown in Table 2.1. Remember that each large watershed has many smaller watersheds nested within it.

Although the term *watershed* can be used as a general one, it is common to use different terms for different-sized watersheds. At the small end (first or second order) would be a *catchment*, and at the larger end would be a *basin*, with the term *watershed* most commonly used in the middle range. However, this usage is not rigidly defined and can vary greatly.

## USGS Hydrologic Units and the Problem with Watersheds

In order to effectively organize water resource data, the USGS has developed a system for dividing the US into *hydrologic units*. This is a hierarchical system, taking advantage of the nested nature of watersheds. At the largest scale, the entire country is divided into 21 *water resource regions*, designated by a two-digit code. Next, each region is divided into *subregions* (total = 222), designated by an additional two digits. Subregions are divided into *accounting units* (total = 352; an additional two digits), which are divided into *cataloging units* (total = 2264; an additional two digits).

TABLE 2.1
Comparison of streams and watersheds of different sizes in the United States

Data from Leopold et al. (1964).

| Stream Order | Number of Streams | Average Length (km) | Total Length (km) | Mean Watershed Area (km²) |
|---|---|---|---|---|
| 1 | 1,570,000 | 1.6 | 2,500,000 | 2.6 |
| 2 | 350,000 | 3.7 | 1,300,000 | 12 |
| 3 | 80,000 | 8.5 | 680,000 | 60 |
| 4 | 18,000 | 19 | 350,000 | 280 |
| 5 | 4,200 | 45 | 190,000 | 1,300 |
| 6 | 950 | 100 | 98,000 | 6,400 |
| 7 | 200 | 240 | 47,000 | 30,000 |
| 8 | 41 | 540 | 22,000 | 140,000 |
| 9 | 8 | 1,300 | 10,000 | 680,000 |
| 10 | 1 | 2,900 | 2,900 | 3,000,000 |

Cataloging units are thus designated by an eight-digit number, referred to as the *hydrologic unit code* or HUC. Cataloging units have an average area of ~1700 km², or approximately the size of the watershed of a fifth-order stream.

It is important to realize, however, that USGS hydrologic units are not necessarily watersheds. Two examples are shown in Figure 2.3. In the first, the hydrologic unit consists of the watershed of the Quinnipiac River (after which the unit is named)—but also includes several nearby small coastal streams that drain to Long Island Sound, such as the Branford River. Each of these streams has its own watershed, which is distinct from that of the Quinnipiac. However, because they are so small, it would be impractical to manage data at the watershed scale, so they are lumped together with the Quinnipiac (the "lumping problem") to form an HUC that is not a true watershed.

In the second example (Figure 2.3 bottom), the hydrologic unit consists of the lower part of the Thames River watershed. It is not a true watershed, since it does not include the two tributaries that join to form the Thames and thus does not include all the land that drains to the outlet. This represents the "splitting problem": the Thames River basin is too large to fit into the desired size of a hydrologic unit, so it must be split into pieces. Note that for the uppermost pieces, this results in a true watershed, but for the lower piece, it results in a hydrologic unit that is not a true watershed.

This illuminates a fundamental truth about watersheds: although the watershed concept is quite flexible, its use is constrained by certain

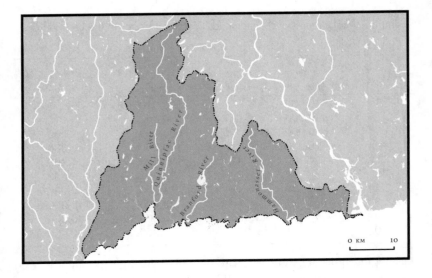

*Figure 2.3.* USGS hydrologic units. Top: HUC 01100004 (Quinnipiac). This HUC is cataloging unit 04 within accounting unit 00 within subregion 10 within region 01. Note that it is not a true watershed: it includes both the relatively large Quinnipiac River and several other smaller rivers that drain to Long Island Sound, such as the Mill, Branford, and Hammonasset rivers. Bottom: HUC 01100003 (Thames), along with HUC 01100002 (Shetucket) and HUC 01100001 (Quinebaug). The Thames HUC is not a true watershed, since it receives water from the other two HUCs shown. Map produced by Stacey Maples, Yale University.

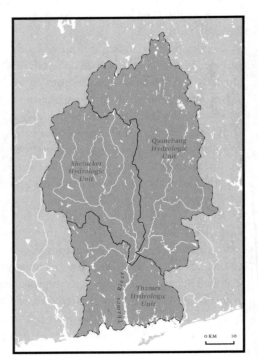

physical realities. This can best be illustrated by examining three contrasting geographical situations. First, if you live in the watershed of a small coastal stream (e.g., Branford River), you live only in a small watershed. In a purely physical sense, your water resource community is only those who share this small watershed with you. In contrast, if you live in the downstream portion of the Thames River basin, you live only in a large watershed. Your water resource community is all those who live in the basin, including those far upstream from you, whose water and land use affects the quantity and quality of water that flows down to you. Last, if you live in a small tributary of the Thames River basin, you have a choice. You are both part of a small watershed and part of the larger watershed. In a purely physical sense, you can choose to ignore downstream users and share the waters of your stream with the other residents of your small watershed. Or you can see yourself as part of the larger Thames River basin and see your water resource community as comprised of all the users of that much larger resource.

The tension between the watershed concept and the use of hydrologic units reflects the physical realities illustrated above. The USGS system is great for dividing up the country into evenly sized units—but the reality is that not all of the country can be divided up that way.[5] The USGS generally does not refer to these units as watersheds, but other agencies are not as careful. In particular, the EPA "surf your watershed" Web site provides water quality information on your "watershed"—but is really organized around hydrologic units. This is a significant error, because as a fundamental physical fact, those who live in the Branford River watershed (for example) are not directly affected by water quantity and quality in the Quinnipiac River basin—yet when they look up their zip code at the EPA Web site, they will find that they are part of the "Quinnipiac watershed."

## 2.3. Introduction to Water Quality

To understand and manage water resources, it is critical to think about the *quality* of water as well as its quantity. Poor water quality degrades aquatic ecosystems and limits human uses of water. In this section, the different types of pollutants are introduced and a brief summary of the US Clean Water Act's approach to controlling them is introduced.

---

5. A rough estimate of how much of the US can be divided into watersheds of the "right" size (the size of a cataloging unit, approximately fifth order) can be obtained from the data in Table 2.1. If there are ~4200 fifth-order rivers with average watershed size 1300 km², then 5.6 million km², or 58% of the US, is found in a fifth-order watershed. The rest is part of watersheds that are either too small or too large.

Concentrations in the environment of the pollutants discussed below can be expressed in a sometimes-bewildering variety of different ways. Box 2.1 provides some guidance to understanding different units and terms.

## Types of Pollutants

Temperature. Anthropogenic temperature deviations—especially increases—can affect habitat suitability in many streams, especially for cold-water fish. The main sources of temperature increases include the discharge of cooling water from power plants, the runoff of rainwater over hot impervious surfaces (e.g., parking lots), and the removal of streamside vegetation and the shading that it provides.

Conductivity. Electrical conductivity (the ability of water to conduct a current) is a measure of the dissolved material in the water. Conductivity may increase either from natural sources (e.g., mineral weathering in soils) or from human impacts (e.g., sewage discharge).

pH. Natural waters have a balance between acids and bases, resulting in a pH that is usually close to neutral (generally between 5 and 9). Discharges of highly acidic (low pH) or basic (high pH) effluents—usually from industrial processes—can lead to serious toxicity to fish and other organisms, as well as damage to water infrastructure.

Turbidity, total suspended solids (TSS), suspended particulate matter (SPM). These are different measures of sediment suspended in the water column. Suspended sediment levels vary naturally from stream to stream and over time (especially during storm events), but many human actions can increase sediment loads, including discharge of municipal or industrial effluents, as well as erosion of poorly protected agricultural or urban soils. In addition, dams can trap sediments and lead to unnaturally clear water downstream. Turbidity is most commonly measured in *nephelometric turbidity units* (NTU) by measuring the scattering of light by particles in the water. In contrast, TSS and SPM are mass measurements (mg/L) of the amount of material captured by filters.

Dissolved oxygen (DO). An adequate level of oxygen dissolved in the water column is vital for most stream organisms. Oxygen concentrations in a water body are controlled by both physical and biological processes.

Physically, aeration of the water (good contact between the water and

## Box 2.1. Ways to Describe Amounts of Pollutants

The amount of a pollutant in a water body is often expressed as a *concentration* (mass per unit volume), for example, 1 mg/L of total N. Since a liter of freshwater has a mass of 1 kg, a concentration of 1 mg/L is equivalent to 1 mg/kg or 1 ppm (part per million), while 1 μg/L is equivalent to 1 ppb (part per billion). For nutrients such as $NO_3^-$, concentrations can include either the mass of the entire compound (e.g., 1 mg $NO_3^-$/L) or just the mass of the relevant atom (e.g., 1 mg $NO_3^-$–N/L, which should be understood as *1 mg of N in the form of $NO_3^-$ per liter of water*). Concentrations can also be expressed in molar units (e.g., 1 μmole $NO_3^-$/L), which can be converted to mass-based concentrations using the molecular weight (e.g., 1 mole of $NO_3^-$ has a total mass of 62 g and a N mass of 14 g).

In a river, when water flow is known, the level of the pollutant can also be described as a *flux* (the amount of the pollutant moving past a point per unit time) simply by multiplying the concentration by the flow, for example, 1 mg/L x 1 ft³/sec = 0.102 kg/hr.

In addition, we often normalize the flux by the watershed area to calculate a *yield*, for example, 1 kg TN ha⁻¹ yr⁻¹. This is meant to express the average amount of nitrogen that is flushed from each unit of land into the river over the course of a year and allows direct comparisons of watersheds of different sizes.

When concentrations of pollutants vary dramatically over time—for example, over the course of a storm—their "average" concentration is often expressed as an *event mean concentration* (EMC). This is the total flux of the pollutant over the course of the event divided by the total flow of water over the course of the event. Equivalently, you can think of the EMC as a weighted-average concentration, where each sample's concentration is weighted by how much water flow it represents.

Pollutants—especially those that are persistent and bioaccumulative—are often measured in fish and sediment as well as in water. Concentrations in these media are generally expressed as mg of pollutant per kg of fish or sediment (ppm) or as μg of pollutant per kg of fish or sediment (ppb). A lab report should indicate whether the concentration is expressed per unit of wet weight or dry weight.

the air) tends to lead to 100% oxygen saturation, that is, to an oxygen concentration that reflects the physical–chemical equilibrium between oxygen in the air and oxygen in the water. The amount of oxygen dissolved in water at saturation is dependent on temperature and salinity, and ranges from ~6 mg/L at high temperature and high salinity to ~14 mg/L at low temperature and low salinity.

However, the actual oxygen concentration in a water body can differ significantly from this saturation level as a result of biological processes. In particular, *photosynthesis* (production of organic material by algae or plants) can lead to supersaturation (high levels of DO), whereas *respiration* (consumption of organic material by bacteria or animals) can lead to undersaturation (low levels). The classic case of *hypoxia* (DO < 3 mg/L) involves inputs of organic matter, such as sewage or other organic wastes, which are respired by bacteria, resulting in depletion of the oxygen in the water. Hypoxia problems are most likely in water masses that are stagnant or somehow isolated from the atmosphere, and thus have slow rates of re-aeration.

Biochemical oxygen demand (BOD), chemical oxygen demand (COD), total organic carbon (TOC). These three measures all reflect the amount of organic matter in a water sample, particularly in relation to its oxygen-depletion potential. BOD is measured by incubating a water sample in the dark for five days and allowing bacteria to respire the organic matter and, in the process, consume oxygen. The difference in oxygen concentration between the beginning and end of the five-day incubation is the $BOD_5$, expressed in mg $O_2$/L. COD is similar, except that the oxygen consumption is carried out by chemical, instead of biological, oxidation. TOC is a direct measure of the organic matter present in the water sample (expressed as the amount of carbon, the main constituent of organic compounds). TOC is generally positively related to BOD and COD, but the exact relationship differs from sample to sample, because not all the TOC is available for oxidation.

Nutrients. In order to grow, plants and algae need certain amounts of inorganic nutrients, especially nitrogen (N) and phosphorus (P). The presence of elevated levels of these nutrients in a water body—sometimes referred to as *eutrophication*—can lead to abnormally high rates of photosynthesis, which can have several negative effects, including harmful algal blooms and hypoxia (caused by respiration of the elevated organic matter supply). Human activities—including agriculture, fossil fuel combustion, and sewage discharges—have greatly accelerated the natural cycles of N and P, and dramatically increased the availability of these nutrients in many water bodies. The most important forms of these elements are the following:

- DIN (dissolved inorganic N): includes all the forms that are available for plants to use, primarily $NH_4^+$ (ammonium), $NH_3$ (ammonia), and $NO_3^-$ (nitrate)
- TN (total N): includes both DIN and organic N; the latter is mostly not available to plants but instead reflects N that has already been incorporated into organic molecules through photosynthesis
- DIP (dissolved inorganic P): primarily $PO_4^{3-}$ (orthophosphate)
- TP (total P): includes both DIP and organic P

Bacterial indicators. Human and animal waste contains many pathogens that can cause disease to people, as we discuss in Chapter 9. To identify the possible presence of pathogens in water, we use several *indicators*: bacterial groups that are not in themselves harmful, but that are relatively easy to measure and that tend to *co-occur* with pathogens. When concentrations of these indicators go above established thresholds, we consider the water potentially contaminated with pathogens and unsafe to use. Different indicators are used for different applications, for example, drinking water, swimming, and so on. The most important indicator groups are the following:

- Total coliform: the most general indicator, used to indicate *possible* water quality problems in drinking water and to require more specific testing
- Fecal coliform: a subset of total coliform that is a bit more specific for human/animal waste, used in some countries for testing of both drinking water and bathing beaches
- *Escherichia coli (E. coli)*: a subset of fecal coliform, this is a specific bacterium that is well correlated with the presence of pathogen-related illness and is used in the US as the primary indicator for freshwater bathing beaches
- Enterococci: used in the US as the primary indicator for saltwater bathing beaches

Metals and organic contaminants. These are sometimes referred to as *toxic micropollutants*, since they are toxic compounds generally found at concentrations much lower than the *conventional* pollutants discussed above.

Metals—such as lead, mercury, and silver—are, of course, naturally occurring, but they are used in large amounts in a variety of human activities (e.g., the addition of lead to gasoline, the use of mercury in the industrial manufacture of chlorine, and the use of silver salts in photography) and often end up being released into waterways. Organic pollutants include both petroleum compounds, such as benzene and other gasoline components, and synthetic compounds, such as chlorinated pesticides. While some organic pollutants (e.g., benzene) break down fairly rapidly in the environment, others (e.g., PCBs) are *persistent* for long time periods. There are almost 100,000 different synthetic compounds in daily use in industrialized societies, and we don't have enough information on the environmental fate and toxicity of many of these. Both metals and many organic pollutants tend to accumulate in sediment (through a process known as *sorption*) and in organisms (*bioaccumulation*), and the main pathway of human exposure is often from consumption of fish and other animal products.

Emerging contaminants. We have recently become aware of additional classes of organic pollutants, beyond the traditional groups of PAHs, pesticides, PCBs, and so on. These emerging contaminants include pharmaceuticals and personal care products, as well as a group of brominated flame retardants known as PBDEs.

The emerging contaminants are only the most recent example of our evolving understanding of pollution. The history of water pollution monitoring is one of continuous increases in the number of different pollutants being monitored. Again and again, we have become aware of a new type of potential chemical threat, started to look for it in our rivers and lakes, and indeed found it. Sometimes it seems that the more we look, the more we find, and that our ability to protect human health and the environment is limited by our lack of foresight. Is there a way to move to a more *precautionary* approach, in which synthetic chemicals are tested for their harmful effects *before* we start using them?

## The Clean Water Act and Pollution Permits

The US Clean Water Act (CWA), passed in 1972, had lofty goals: to "restore and maintain the chemical, physical, and biological integrity of the Nation's waters." In order to do this, it needed to address the huge fluxes of pollutants into water bodies, from both *point sources*—direct discharges of effluents from pipes into a water body; and *nonpoint sources*—diffuse runoff of pollution from a large area.

Point sources were addressed primarily through a system of pollution permits, known as the National Pollutant Discharge Elimination System, or NPDES. Section 402 of the CWA requires point sources to obtain permits from the EPA (or from the state, for the 46 states that have been authorized to run their own permitting program under EPA supervision). These permits specify acceptable levels of pollutants in the effluent, as well as monitoring and reporting requirements.

The initial permitted effluent levels were technology based; that is, they were derived for each type of point source based on the technology available to treat that type of effluent. For example, for municipal sewage treatment plants, permits typically require a 95% removal of suspended solids and BOD; this is based on the capacity of secondary treatment to achieve these removals. This uniform application of treatment technology greatly simplified the writing of permits and led to the successful implementation of secondary sewage treatment in almost all cities in the US. However, these technology-based permit levels are not adequately protective of water quality in all cases, and they are increasingly being supplemented with water quality–based permit levels (see section 8.7).

Enforcement of NPDES permits is weaker than it should be (Duhigg 2009a). Of the ~6500 "major facilities" covered by NPDES permits, 24% were in "significant non-compliance" in 2008 (EPA 2009a).[6] Most of these violations resulted in some type of state or EPA enforcement action, but most of the enforcement was informal and the fines paid were relatively small (total $4.3 million; EPA 2009a). The compliance and enforcement record is slightly worse for the ~39,000 "non-majors" with NPDES permits (EPA 2008). Interactive databases of facility compliance and enforcement are available through both the EPA (www.epa-echo .gov/echo/index.html) and the *New York Times* (projects.nytimes.com/ toxic-waters/polluters).

Controlling nonpoint sources has proven even more difficult than regulating point sources. The CWA approach to protecting water quality is discussed further in section 8.7.

---

6. Major facilities include municipal sewage treatment plants with discharges of at least 1 million gallons per day, as well as industrial facilities that meet certain criteria. Significant noncompliance is calculated by the EPA based on the duration, severity, and type of permit violations.

## 2.4. Introduction to Fluvial Geomorphology and Aquatic Ecology

We end this chapter by very briefly examining two additional facets of water resource science. The disciplines of geomorphology and aquatic ecology are critical for understanding aquatic ecosystems such as rivers and lakes. Here we restrict ourselves to a brief discussion of the concepts from these disciplines that are most fundamental and most relevant to our analysis of water resources.

Geomorphology is the study of the shape of the earth's surface and how it changes over time. Of interest to us here is fluvial geomorphology, which examines the processes that shape river channels and their floodplains. A key insight of this field is that the shape of a river channel reflects a balance between sediment sources, on the one hand, and water velocities capable of moving those sediments, on the other.[7] Four factors are relevant to this balance:

- Sediment inputs: Erosion from the watershed delivers sediment to the river. All other things being equal, an increase in the sediment supply will tend to cause the river channel to aggrade (accumulate sediment).
- Sediment size: Larger particles are harder for water to move. An increase in sediment grain size will tend to lead to aggradation.
- Slope: The slope of the river channel (the loss in elevation per unit of distance along the river) affects water velocity; for a given flow of water (and a given riverbed composition), higher slope means higher velocities. Water velocity, in turn, affects sediment movement, since higher water velocities can move more (and larger) sediment. Thus, all other things being equal, an increase in slope will tend to cause the river channel to degrade (lose sediment and cut into its bed or banks).
- Flow: Higher flows can carry more sediment. Streams often accumulate sediment at low flow and then flush it out at higher flows. The flows that are most important in shaping the channel are the "bank-full" or "channel-forming" flows, which for many rivers occur roughly every two years. A sustained increase in flows will tend to lead to channel degradation.

---

7. This is strictly true only for alluvial rivers (those whose beds consist of unconsolidated sediment), not for rivers that flow over bedrock.

The sediment balance described above changes systematically as one moves from small headwater streams to large mainstem rivers. In many regions, small streams tend to have steep slopes and high velocities, leading to sediment erosion and transport, while larger rivers have gentler slopes and lower velocities, leading to sediment deposition. Beyond this simplistic description, many other factors can affect channel shape.

Several geomorphic classification schemes have been developed for rivers. The best known is probably Rosgen's (1996), which classifies rivers based on their slope, channel material, sinuosity (the degree to which they meander), width–to–depth ratio, and entrenchment ratio (a measure of the relationship between the channel and the floodplain).

*Aquatic ecology* focuses on the organisms that live in a water body, particularly their interactions with each other and with the abiotic environment. One fundamental area of ecology is the cycling of energy and materials, captured in the concept of the *food chain*. The base of the food chain consists of *primary producers* (plants, algae, certain types of bacteria), who use the energy of sunlight to carry out photosynthesis and produce organic matter. This first *trophic level* is then fed upon by herbivores (a second trophic level), who are fed upon by carnivores (third trophic level) and so on. In each trophic transfer, some energy is transferred up to the higher level, but most of the energy is lost through respiration. This results in a *trophic pyramid*, in which successively higher trophic levels receive less and less energy. At each level, there is also some energy that goes to decomposers (bacteria).

In streams, the primary producers are generally periphyton (algae attached to the bottom), macrophytes (plants or mosses growing on the bottom), or phytoplankton (algae growing in the water column). A variety of invertebrates fill the herbivore niche, and fish dominate higher trophic levels.

A number of models have been constructed to describe differences in the ecology of small and larger streams. Perhaps the most popular is the *river continuum concept* (Vannote et al. 1980), which emphasizes the gradient in energy sources: from small, shaded streams that are driven by external inputs of coarse organic matter (e.g., leaves); to midsized streams where most organic matter comes from photosynthesis by macrophytes and periphyton; to large, turbid rivers where the main sources of organic material are transport from upstream and photosynthesis by phytoplankton near the water surface. These differences in energy sources then result in differences in the structure of the biotic community, such as the types of invertebrates or fish that dominate.

## 2.5. Conclusion

This chapter provides the foundation of biophysical knowledge for the rest of the book. You should now have a good understanding of some basic concepts: the hydrologic cycle, linkages between surface and groundwater, the watershed as a unit of analysis, the variation in hydrologic properties as a function of scale, ways of expressing pollutant concentrations, the different types of pollutants and their sources, point source permits, the factors that shape river channels, and the ways that energy cycles through food chains. Of course, there is much more to learn about each of these topics, and you are encouraged to delve deeper using the books listed at the beginning of the Recommended Readings section. The next chapter moves to looking at water more directly as a resource for human use, by examining water availability and use around the world.

# 3

## Supply and Demand: Water Availability and Water Use

How much water is available in different parts of the world? How much are people using? This chapter provides an overview of water availability, on the one hand, and water use, on the other. Throughout the chapter, we look closely at water data at different scales, ranging from global to local.

### 3.1. Global Water Resources

#### Stocks and Flows

Discussions of global water resources often start with the amount of water in various environmental compartments, such as oceans, glaciers, groundwater, lakes, and rivers (Table 3.1). However, this emphasis on *inventories* or *stocks* (the static amount of water in a given environmental compartment) is misplaced, because water—unlike, say, copper or oil—is a *renewable resource*. The water in rivers is continually replenished through the hydrologic cycle. If we want to achieve sustainable water use, our emphasis should be on the *fluxes* or *flows* of this resource (the movement of water from one compartment to another). Knowing the amount of water in an aquifer or a lake is useful, but if that water is not being replenished (or if we are using it more rapidly than it is being replenished), then human withdrawal of that water is, in the long run, unsustainable.[1]

---

1. I am using the term *sustainability* here in the purely physical sense, to refer to using water in a manner that does not change the amount of water available to future generations. Sustainability is a complex concept with many different definitions.

TABLE 3.1

Water stocks on Earth

Data from Shiklomanov and Rodda (2003).

| Water Source | Volume ($10^3$ $km^3$) | Percentage of Global Reserves (%) | |
|---|---|---|---|
| | | Percent of Total | Percent of Freshwater |
| oceans | 1,338,000 | 96.5 | — |
| glaciers and permanent snow | 24,060 | 1.74 | 68.7 |
| saline groundwater | 12,870 | 0.94 | — |
| fresh groundwater | 10,530 | 0.76 | 30.1 |
| ice in permafrost | 300 | 0.022 | 0.86 |
| fresh lakes | 91.0 | 0.007 | 0.26 |
| saline lakes | 85.4 | 0.006 | — |
| soil moisture | 16.5 | 0.001 | 0.05 |
| atmosphere | 12.9 | 0.001 | 0.04 |
| swamp water | 11.5 | 0.0008 | 0.03 |
| rivers | 2.12 | 0.0002 | 0.006 |
| biological water | 1.12 | 0.0001 | 0.0003 |

A note on units: Stocks are expressed in volume units, while flows are expressed as volume per unit time. Many different units are used to describe water volumes, including $m^3$, $ft^3$, gallons, acre-feet (the volume of water that will cover an acre of land with a foot of water), and so on. Some of the most important conversions are shown in Box 3.1.

A focus on flows leads to an analysis of the *global hydrologic cycle* (Figure 3.1). For the global ocean, loss of water through evaporation is greater than inputs of water from precipitation. The difference is transported as atmospheric water vapor to the continents, where it contributes to an excess of precipitation over evaporation. This excess then flows as runoff back to the oceans, completing the cycle. It is this flow—the runoff back to the ocean of freshwater originally evaporated from the ocean—that is the renewable "blue water" resource that is available for human use year

---

### Box 3.1. Unit Conversions for Water Resources

- volume: 1 $km^3$ = 1 billion $m^3$ (BCM) = 1000 million $m^3$ (MCM) = 0.8107 million acre-feet (MAF) = 264.2 billion gallons
- flow: 1 cusec = 1 $ft^3$/sec (cfs) = 28.32 L/sec = 0.02832 $m^3$/sec = 0.6463 million gallon/day (MGD) = 0.8936 MCM/yr
- per capita use: 1 liter person$^{-1}$ day$^{-1}$ (Lpcd) = 0.2642 gallon person$^{-1}$ day$^{-1}$ = 0.3652 $m^3$ person$^{-1}$ year$^{-1}$.

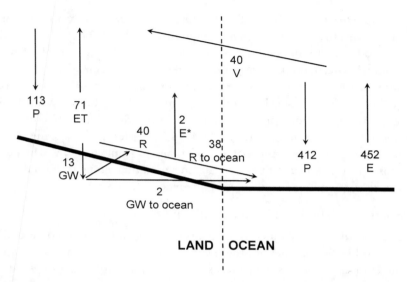

*Figure 3.1.* Global hydrologic cycle in an average year. Fluxes shown are in 1000 km³/yr. P = precipitation; E = evaporation; ET = evapotranspiration; E★ = evaporation after being included in streamflow (~1000 km³/yr of flow to endorheic basins and ~1000 km³/yr of river evaporation between gauging stations and discharge to the ocean); V = water vapor transport from the atmosphere over oceans to the atmosphere over land; R = surface runoff; GW = groundwater recharge (~13,000 km³/yr, of which ~2000 km³/yr discharges directly to the ocean and the remainder discharges to surface water; the latter is included in the 38,000 km³/yr of surface runoff to the ocean). Discharge from Antarctica is not included. The total global blue water resource is estimated to be ~42,000 km³/yr (including flow to the oceans, flow to endorheic basins, and streamflow that evaporates before reaching the oceans). Modified from Falkenmark and Rockstrom (2004) using data from Dai and Trenberth (2002), Shiklomanov and Rodda (2003), and Döll and Fiedler (2008).

after year. Estimates of this resource vary, but it is approximately 42,000 km³ of water per year on a global basis. This includes the annual flow of all of the world's rivers, as well as groundwater that discharges to the oceans.[2] Note that this total of 42,000 km³/yr actually includes a small amount of water (about 2000 km³/yr) that never reaches the ocean—either because it discharges to an endorheic lake (such as the Great Salt Lake) or because

2. Groundwater that discharges to rivers (as opposed to directly to the ocean) is already counted as part of river flow. Thus, recharge of groundwater that ultimately discharges to a river is part of the renewable water resource, but it is usually simpler to account for it as part of river flow.

it evaporates from a river before reaching the ocean (mostly a factor for large rivers that flow through arid regions, such as the Nile).

In addition to the *blue water* or *runoff* resource, hydroecologists have recently drawn attention to the *green water* resource (Falkenmark and Rockstrom 2004). Green water is the precipitation that never makes its way to surface water, but instead evaporates or transpires from the soil. While this water is not available for capture and use in the same way as blue water, it is still an important resource for humans and terrestrial ecosystems. The transpiration of this green water is what allows plants to grow, whether those plants are natural forests or rainfed agricultural crops. (Irrigated agriculture, in contrast, involves using the blue water resource.)

## Spatial Variability in Water Availability

The global water resource is, of course, not evenly distributed around the world—and neither are the people who want to use that water. In this section, we discuss the patterns of spatial variability in the three important components of the global hydrologic cycle: precipitation, evapotranspiration, and runoff.[3]

The spatial variability in average annual precipitation is shown in Figure 3.2. Atmospheric circulation patterns result in some parts of the world receiving very little rainfall (e.g., much of northern and Southern Africa, the Middle East, central Asia, interior Australia, the western United States, and parts of South America), while other areas have very high annual precipitation rates (e.g., parts of central Africa, parts of India, Southeast Asia, Central America, and the Amazon region). One important feature in Figure 3.2 is the Inter-Tropical Convergence Zone (ITCZ), a band near the equator where surface winds converge, leading to rising air, adiabatic cooling, and high precipitation rates. Conversely, at about 30 degrees north and south latitude, sinking air masses lead to dry conditions.

In addition to precipitation, we also need to account for evapotranspiration in determining net water availability. The Climate Moisture Index (CMI; Willmott and Feddema 1992) takes into account both precipitation (P) and potential evapotranspiration (PET, i.e., evaporative demand, not actual ET):

---

3. Much of the discussion below is based on spatially explicit global datasets with a spatial resolution of 0.5 degree (about 55 × 55 km at the equator) created by researchers at the University of New Hampshire. Many of these datasets were created for the UN's Second World Water Development Report and are available for download (wwdrii.sr.unh.edu/).

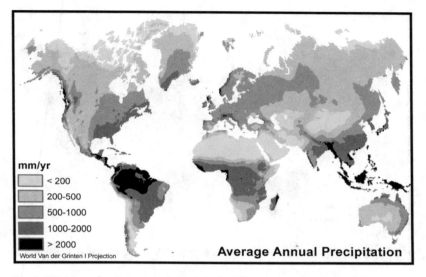

*Figure 3.2.* Map of average annual precipitation. Data source: University of New Hampshire Water Systems Analysis Group (wwdrii.sr.unh.edu/). Map produced by Stacey Maples, Yale University.

$$\text{CMI} = 1-(\text{PET}/P), \text{ for PET} \leq P$$
$$\text{CMI} = (P/\text{PET})-1, \text{ for PET} > P$$

This produces an index that ranges from 1 (very humid) to −1 (very arid). Where CMI is negative (evaporative demand is greater than annual precipitation), we expect little net water availability; these are the great deserts and arid zones of the world. Where CMI is highly positive, large amounts of runoff are generated and great river systems form. Figure 3.3 illustrates the global patterns of CMI.

Actual ET is shown in Figure 3.4 and is quite different from PET. In particular, it is low in very arid regions due to the lack of water available for evaporation.

Runoff generation (precipitation minus ET) is shown in Figure 3.5. Note that there are areas with very little *generation* of runoff that nonetheless have substantial water resources—in the form of large rivers that run through them.[4] Egypt, which receives practically no precipitation but has the Nile flowing through it, presents a dramatic example of this phenomenon.

---

4. Also available is a simulated global river network based on runoff generation and topography; see www.grdc.sr.unh.edu/html/Stn.html.

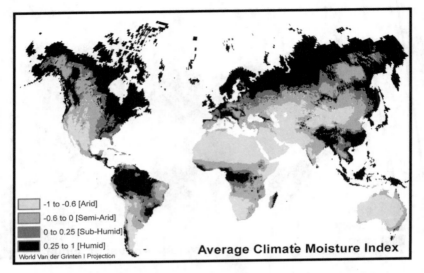

Figure 3.3. Map of the Climate Moisture Index. Data source: University of New Hampshire Water Systems Analysis Group (wwdrii.sr.unh.edu/). Map produced by Stacey Maples, Yale University.

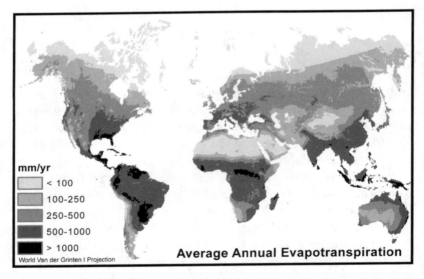

Figure 3.4. Map of average annual evapotranspiration. Data source: University of New Hampshire Water Systems Analysis Group (wwdrii.sr.unh.edu/). Map produced by Stacey Maples, Yale University.

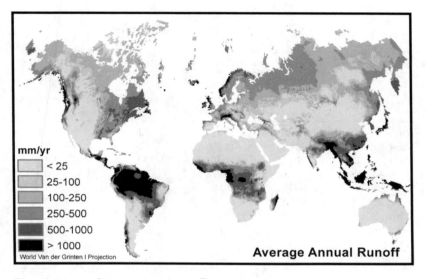

*Figure 3.5.* Map of average annual runoff generation. Data source: University of New Hampshire Water Systems Analysis Group (wwdrii.sr.unh.edu/). Map produced by Stacey Maples, Yale University.

Using runoff data combined with population data, we can calculate *per capita water availability* for different areas, as shown in Table 3.2. At the continental scale, Africa has the lowest runoff per unit of land, while Asia and Europe have the lowest runoff per capita.

When dealing with any type of spatial water resource data, it is crucial to pay close attention to the scale of the analysis. Whether you are dealing with runoff generation or per capita water availability or other measures, you will find that as the scale gets smaller, the measures become more extreme (both small and large). This is reflected in its simplest form in Table 3.2, where the continents vary in their water availability from the global average, with some continents higher and others lower. As we go to finer scale and look *within* a continent, we find areas of higher and lower water availability; that variability gets averaged out at the scale of the continent. As an extreme example, it is likely that the per capita water availability in the room you are sitting in is very low (perhaps a glass of water per person). This is obviously not an appropriate scale, since you can go into another room and get more water. But what *is* the right scale? This is a difficult question. For now, we turn to the country scale.

TABLE 3.2
Renewable water resources by continent

Data from Shiklomanov and Rodda (2003). Note that the units for runoff per unit area are volume area⁻¹ time⁻¹, which can also be expressed as depth/time.

| | Average Annual Runoff (km³/yr) | Coefficient of Variation of Annual Runoff | Runoff per Unit Area (mm/yr) | Runoff per Capita (m³ person⁻¹ yr⁻¹) |
|---|---|---|---|---|
| globe | 42,757 | 0.02 | 317 | 7,600 |
| Africa | 4,047 | 0.10 | 134 | 5,720 |
| Asia | 13,510 | 0.06 | 311 | 3,920 |
| Australia and Oceania | 2,400 | 0.10 | 268 | 83,600 |
| Europe | 2,900 | 0.10 | 277 | 4,240 |
| North America | 7,870 | 0.10 | 324 | 17,400 |
| South America | 12,030 | 0.07 | 672 | 38,300 |

## Water Availability at the Country Scale

In Chapter 2 we saw that the watershed is, from a biophysical perspective, the natural unit for water budgets. At the same time, water is often managed not at the watershed scale, but at the scale of individual countries or states, and thus we do need to be able to do our water accounting at those scales as well. In this section, we first discuss the methodology of water budgets at the country scale and then address the question of how much water different countries have.

Current data on country-level water availability and withdrawals are compiled by the Food and Agricultural Organization of the United Nations (FAO), in their AQUASTAT database (www.fao.org/nr/water/aquastat/dbase/index.stm). An important caveat (common to this and many other global data sources): the database consists of data collected from individual countries, and the timeliness and quality of the data are quite variable.

Figure 3.6 illustrates the methodology used by the FAO to do accounting of freshwater availability at the country scale. Some fraction of *precipitation* that falls within the country becomes *surface water* and *groundwater* flow. These two combined make up the *internal renewable water resources* (IRWR) of the country, although we must be careful to subtract the *overlap* between surface and groundwater in order to avoid double counting. The overlap is generally high in humid regions, where groundwater typically discharges to rivers as baseflow; in this situation, groundwater should not be counted as an additional resource above surface water flow.

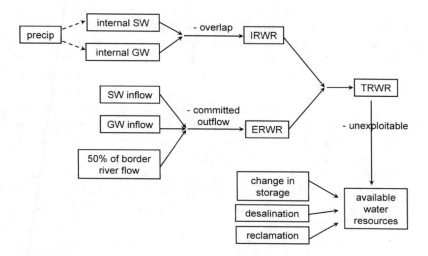

*Figure 3.6.* Schematic illustration of country-level water accounting used in the AQUASTAT database. SW = surface water; GW = groundwater; IRWR = internal renewable water resources; ERWR = external renewable water resources; TRWR = total renewable water resources.

The overlap is typically low in coastal regions, where groundwater may discharge directly to the ocean, and in arid and semiarid regions, where groundwater may discharge to an evaporation sink. Remember, also, that we are not counting fossil groundwater in the renewable water resource.

For countries that share a river basin with other countries, we need to account also for *external renewable water resources* (ERWR). We will discuss international water sharing further in Chapter 13, but for now, we simply need to know how much water is flowing into our country from upstream and how much needs to flow out of our country for use by downstream countries. The inflow from upstream countries is usually thought of as river flow (surface water), but groundwater may also flow across country boundaries, although this is generally much harder to quantify, especially when it is not in the interests of one of the countries to do so. In addition, we need to account for the special—but relatively common—case in which a river makes up the border between two countries. In this case, the accounting rule used by the FAO is to assign each country half of the flow in the shared resource.

Calculations of ERWR also take into account the flow that must be released to downstream countries ("committed outflow"), since this water is not available for use by the country under discussion. Exactly what is meant by "must be released" is an area of potential ambiguity in this definition, since it could mean the amount required by treaty, the amount expected by custom, or simply the amount currently being released.

We should note that while IRWR is a purely physical quantity (and will change only in response to changes in climate), ERWR is defined by the behavior of the countries that are upstream and downstream of the country under consideration. In particular, a country's ERWR may decrease if upstream countries start using more water and thus leaving less in the river, or if the country signs a treaty that requires it to increase the amount that it leaves in the river for downstream countries.

ERWR and IRWR combine to give a country's *total renewable water resources* (TRWR). A useful measure of the country's dependence on external water sources is the *dependency ratio*, which is simply the ratio of ERWR to TRWR.[5]

In terms of calculating the water actually available for use (*available water resources*), we might want to make several adjustments to TRWR (Figure 3.6). First, we need to acknowledge that some of the water is not really exploitable for human use, because it is too remote or runs off in large floods.[6] Of the global TRWR of ~42,000 km$^3$ yr$^{-1}$, it is estimated that ~9000–14,000 km$^3$ yr$^{-1}$ are exploitable (Seckler et al. 1998), although this should be considered a very rough estimate. Estimates of country-level exploitable water resources are available for only a few countries. In addition, we may want to include in the available water resources two nontraditional water sources—desalination and wastewater reclamation—both of which will be discussed further in Chapter 7. Last, we may want to include the change in stocks in a country. Thus, for example, many countries are reducing the amount of water stored in aquifers by withdrawing groundwater at a rate faster than it is recharged. This groundwater use is not included as part of TRWR but is certainly an available water resource, at least in the short term.

In practice, because of data limitations and because of the desire to focus on renewable water resources, it is TRWR, rather than available water resources, that is generally used as the measure of a country's water resources.

---

5. Technically, the dependency ratio is often defined using gross, rather than net, inflow, that is, without accounting for committed outflow.

6. Note that the definition of what is "exploitable" varies with time, as technological, economic, political, and social constraints change. As a generalization, groundwater flow and "regular" surface water flow (i.e., baseflow) are usually considered exploitable, while flood flows in rivers may or may not be exploitable, depending on the feasibility of constructing capture and storage facilities (usually dams).

The type of analysis represented by Figure 3.6 can be carried out at a variety of scales, ranging from a small watershed or municipality to a large country or region. TRWR will obviously change as the scale of analysis changes. What may be less obvious is that TRWR can actually *decrease* as we move to larger scale. This typically occurs when a large river flows through an arid area, where the water lost to evaporation is greater than the water added by runoff. For example, as the Nile flows through Sudan, it loses a tremendous amount of water to evaporation in the Sudd swamps. Thus the TRWR of the Nile watershed is actually greatest where it leaves the highlands and decreases as it flows downstream.

Finally, we should note that the entire discussion above focuses on "blue water," that is, surface and groundwater resources, but does not include "green water," soil water that evaporates or transpires. A simple measure of country-level green water availability is the amount of precipitation that does not become blue water. However, the significance of this green water as an addition to the country's water budget will vary tremendously, depending on the timing and location of that green water flow, as well as the types of ecosystems that it supports (e.g., evaporation from lakes versus transpiration from rain-fed agricultural fields).

In any case, the FAO uses the methodology of Figure 3.6 to calculate per capita water availability for every country. Not surprisingly, countries vary tremendously in their water endowment. The most water-rich country is Greenland, with over 10,000,000 $m^3$ person$^{-1}$ yr$^{-1}$, while the most water-scarce country is Kuwait, with 7 $m^3$ person$^{-1}$ yr$^{-1}$. For large countries, like China, the average can mask large variation within the country. We return to water availability at the country and subcountry scales when we discuss scarcity in Chapter 5.

## Temporal Variability in Water Availability

In the FAO database (and most other sources), water availability is expressed as an annual average. However, it is crucial to consider the *variability* around this average at different temporal scales. *Seasonality* can be an important factor in water availability. Some regions (e.g., northeastern US) experience uniform precipitation throughout the year, while other regions (e.g., India) experience a strong seasonal cycle in precipitation.[7] As noted in Chapter 2, even where precipitation is uniform, seasonal differences in ET can lead to large differences in runoff (lower runoff in warmer months). In

---

7. This is illustrated beautifully by a global-scale animation of monthly rainfall from 1979 to 1999, available at precip.gsfc.nasa.gov/gifs/gpcp_rain_79-99.qt.

regions with extreme seasonality (typically associated with monsoon rains), almost all of the precipitation may occur in a few weeks; rivers may flood in the wet season and not flow at all in the dry season.[8]

Highly seasonal water availability poses a real challenge for water managers. Brown and Lall (2006) have shown that intra-annual variability in rainfall is a good predictor of a country's level of development: countries with a higher coefficient of variation (CV) of monthly rainfall tend to have lower Gross Domestic Products (GDPs), perhaps in part because of the development constraints posed by low water availability in the dry season.

In addition to regular seasonal patterns, water resource managers must be aware of *interannual variability*. We know that just because average rainfall for July is 100 mm does not mean that it will actually rain 100 mm in a given July. One measure of interannual variability at the continental scale—the CV of annual runoff—can be seen in Table 3.2, which shows that the continents with the most runoff (South America and Asia) have the least variability (CV = 6%). Of course, variability at the global scale is even smaller (2%), since it is unlikely that the entire world will simultaneously experience significant deviations from the norm. Even weather patterns that bring precipitation changes over a significant fraction of the globe, such as El Niño, tend to bring unusually wet conditions to some areas while bringing unusually dry conditions to others.

Another commonly used measure of variability in runoff is the *flow duration curve*. This graphical representation of flow variability is constructed from whatever flow data are available for a given river, generally using daily or monthly flow records. Different percentiles of flow are calculated and plotted, as shown in Figure 3.7. In this graph, the y-axis represents different flows and the x-axis represents how often those flows are exceeded. Low flows (bottom right of the curve in Figure 3.7) are exceeded most of the time, while high flows (top left) are exceeded only a small percentage of the time. Flow duration curves can be used for a variety of analyses. Some commonly used parameters that can be extracted from these plots include the following:

- Q90: the flow that is exceeded 90% of the time; this is a measure of typical—but not extreme—low flows for a given river.

---

8. For runoff generation globally with a monthly time step, see www.compos iterunoff.sr.unh.edu/html/Runoff/monthly.html. For patterns in streamflow in different river basins, see www.grdc.sr.unh.edu/html/Stn.html and select a station of interest.

- Q99 (see Figure 3.7): the flow that is exceeded 99% of the time; this is often used as a measure of extreme low flows or drought (see section 5.4).
- Q95/Q5: the ratio of the 95% exceedance flow to the 5% exceedance flow; this ratio is a measure of *how variable* flows are for a given river.

We return to the issue of extremes in water availability—floods and droughts—in Chapters 4 and 5, respectively.

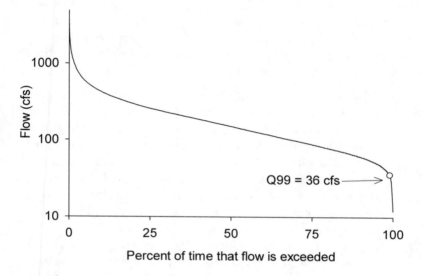

Figure 3.7. Flow duration curve for the Quinnipiac River, Connecticut, based on daily flows for water years 1931–2005 (data source: USGS 2009, station 01196500). Open circle indicates the Q99, the flow exceeded 99% of the time.

## 3.2. Water Use

We now turn to the demand part of the water picture. People use water. But for what purposes? And how much? We first deal with definitional issues and then turn to actual water use data.

### Defining Water Use

The field of water use is plagued with inconsistent and complex terminology. In this section, I try to define terms clearly and use them consistently, but if you read the primary literature, you are likely to encounter some terms being used differently than I use them here.

Most of the data and analysis on water use deal with blue water use, that is, human use of surface or groundwater, rather than green water use, that is, human use of the evaporation or transpiration of soil moisture.

The most basic distinction in blue water use has to do with the location of use. *In-stream* uses do not actually remove water from a river, but rather use it in place. Included in this category are recreational uses (e.g., canoeing), environmental uses (e.g., supporting aquatic life), and most hydropower. In-stream uses are often not well defined and are not included in most water use datasets. In contrast, *off-stream* uses involve withdrawing water from surface or groundwater sources and using it at another location; data on off-stream use are available for most countries, although the completeness and timeliness of the data vary greatly.

Off-stream water uses (water withdrawals) are often classified by *sector*:

- *Domestic (residential)*: Water is used within people's homes for personal hygiene, food preparation, clothes washing, the flushing away of wastes, irrigation of landscape or garden plantings, and so on. Commercial and institutional users are generally included in the domestic category, and small industrial facilities that obtain water from a utility may sometimes be included as well.
- *Agricultural*: Irrigation water is applied to agricultural fields in areas where rainfall is insufficient to grow the desired crops. To a lesser extent, water is used in raising livestock as well.
- *Industrial*: Water is used in practically every industrial process, as a cleaner, coolant, solvent, transport agent, and so on. One large industrial use in many developed countries is the cooling of thermoelectric power plants.

We return to these three sectors in more detail in Chapters 9–11.

It is important to note that the water that we use directly in our homes (domestic water use) is only a small fraction of our total water needs, since every product or piece of food that we use also involves an indirect use of water—the water that it took to make the product or grow the food. This water has been referred to as the *embodied water* or *virtual water* that is implicitly contained in the product.

When water is used for any purpose, some fraction of it may be consumed (*consumptive use* or *depletion*), while the remainder of it will return to the aquatic system (*nonconsumptive use* or *return flow*). Consumptive use consists of four types of activities, which share the characteristic of making the water unavailable for further use:

- *Evapotranspiration*: For example, water that is applied as irrigation water and then transpired by crops is not available for further use, since it ends up as water vapor in the atmosphere.[9] Evapotranspiration is generally the largest type of consumptive use.
- *Incorporation into product*: Water that actually ends up in a product (e.g., agricultural crop, industrial product) is not available for further use; this is usually a small portion of the amount of water used in making the product.
- *Flow to a sink*: When the return flow from a user flows into a sink from which it can no longer be used as freshwater (e.g., the ocean), that is a consumptive use.
- *Pollution*: When the return flow is so polluted that it can't be used by others, that should be considered a consumptive use. In practice, however, polluted return flow is usually considered nonconsumptive, both because the pollutants can (at least in theory) be removed, and because the level of pollution that is required to make the water unusable is dependent on what exactly the downstream use is.

In contrast to consumptive use, return flow is the water that can potentially be used again by downstream users. Examples of uses that are largely nonconsumptive are household uses for washing and industrial uses for cooling. Except in coastal areas, this water flows back into surface or groundwater[10] and can be withdrawn again by other users downstream. A given parcel of water may be withdrawn and used multiple times before being consumed.

In general, agricultural uses tend to have higher consumptive fractions than domestic and industrial uses, since transpiration of water (a consumptive use) is key to crop growth. Note that in-stream uses can also have a consumptive component, despite having no withdrawal of water; in particular, evaporation from reservoirs can lead to large consumptive losses.

---

9. This additional water vapor in the atmosphere can potentially result in additional precipitation within the same basin, which would mean that not all of the evaporated water is truly consumed. However, this only applies when the analysis is being carried out at a large scale in a basin where much of the precipitation is internally generated (e.g., Amazon basin).

10. Return flows to groundwater (e.g., percolation of irrigation water) are hard to measure, and in practice often end up being counted as consumptive use. In theory, however, if these return flows replenish groundwater and allow other users to pump more water, they really should be considered nonconsumptive.

The fact that water can be used multiple times within the same basin is a key feature of water use. While other resources can also be recycled and reused, water is perhaps unique in that so many of its most important uses are nonconsumptive in nature. The distinction between consumptive and nonconsumptive use also has implications for how to define water conservation (Box 3.2).

## Water Use at the Global Scale

With the definitional issues behind us, we now turn to actual water use data to try to answer the key questions about human use of freshwater: How much water are we using? Which sectors are the biggest water users? How has water use changed over the last century? How much do

---

### Box 3.2. Defining Water Conservation

The concept of nonconsumptive use has led to some confusion over what we mean—or should mean—by water conservation. Some have argued that the focus of conservation should be exclusively on reducing consumptive uses, since nonconsumptive uses are not really lost from the system. They point out, for example, that using less water while brushing teeth in an upstream city does not make any more water available to a downstream city (or a downstream ecosystem), since that water was, in any case, returning to the river (as sewage—hopefully treated) and available for reuse downstream. They refer to this type of conservation measure as producing *paper water*, as opposed to real changes in water availability. They argue that conservation efforts should be focused instead on reducing the consumptive use of water (or maximizing the benefit obtained per unit of water consumed).

Gleick (2003) provides what I see as a reasonable response to this argument. He acknowledges the value of distinguishing between consumptive and nonconsumptive uses in water conservation efforts but argues that reducing nonconsumptive uses is still important. In the toothbrushing example, reducing nonconsumptive use in the upstream city—while it does not make any new water available downstream—can reduce the need for developing new sources to serve growth in the upstream city. In addition, it can minimize the impact of withdrawals on local aquatic ecosystems and can lead to higher water quality in the receiving waters downstream. I would also point out that there is a real cost associated with treating and delivering water; wasteful nonconsumptive uses incur that cost just as much as wasteful consumptive uses (in fact, more so, because nonconsumptive uses increase the volume of wastewater that must be treated).

different countries vary in their per capita water use, and is this variability a function of level of development?

To answer these questions, we can draw on two main data sources: Shiklomanov (1999), who compiled historical data (1900–1995) by country (78 countries only), by natural-economic region, by continent, and for the globe as a whole; and FAO's AQUASTAT database, which provides country-level data (1960–2008, but most complete for the year 2000) on water withdrawals (but not consumption or reservoir evaporation). Unfortunately, both of these sources are constrained by the fact that country-level data on water use are incomplete and of variable quality and timeliness. Partly in response to this problem, models such as WaterGAP (Alcamo et al. 2003) have been developed to estimate water use (and availability) for the entire globe at a grid scale (typically with a grid cell size of 0.5 degree).

Data on global withdrawals and consumptive use are given in Figure 3.8, which shows that total global water use grew dramatically over the course of the twentieth century. This was due in part to the growth in world population (by a factor of 3.7) over that time period, but also to a growth in per capita water use (1.9-fold increase in per capita withdrawal, 1.8-fold increase in per capita consumption).

As can be seen in Figure 3.8, agriculture is by far the largest water user globally, although its growth rate over the twentieth century was actually lower in relative terms than for other sectors (see the figure legend for per capita increases over this time period). For the year 2000, agriculture, with its relatively high consumptive fraction, made up 66% of water withdrawn and 84% of water consumed. Of the other categories of use, evaporation from reservoirs (which is entirely a consumptive use) was larger than the water consumed by industrial and domestic uses combined. However, the industrial and domestic sectors withdrew many times more water than they consumed (9 and 7 times as much, respectively). Both industrial and domestic water withdrawals increased dramatically over the twentieth century (4.9-fold increase in per capita withdrawals), though there are signs that industrial water use is leveling off.

It should be noted that while the WaterGAP model agrees with the Shiklomanov/AQUASTAT estimates for water withdrawals, it has much lower estimates of water consumption (total for the year 2000 of 1300 $km^3/yr$ vs. 2240 $km^3/yr$; Döll 2009). This is probably due to two factors: the absence of reservoir evaporation from the WaterGAP consumptive use estimates and the fact that loss of irrigation water to percolation would not be counted by WaterGAP as a consumptive use, while in practice

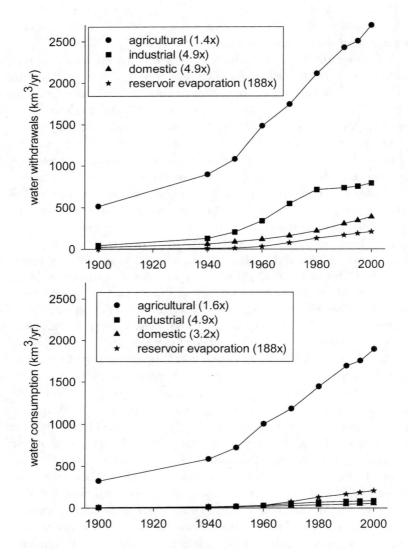

*Figure 3.8.* Global water withdrawals (top) and consumption (bottom) by sector over the twentieth century. The numbers shown in the legend represent the factor by which that use grew *per capita* over the twentieth century. Data for 1900–1995 are from Shiklomanov (1999); data for year 2000 reservoir evaporation are from Shiklomanov (1999)'s projection; data for year 2000 agricultural, industrial, and domestic withdrawals are from AQUASTAT (2009); data for year 2000 agricultural, industrial, and domestic consumption are calculated by the author based on AQUASTAT withdrawal data and a consumptive fraction for each sector estimated from the trend in consumptive fractions in Shiklomanov's data.

much of this water would be considered consumptive by Shiklomanov and AQUASTAT.

As mentioned above, data on green water use are even more uncertain than for blue water use. Using different methods and definitions, four groups have calculated agricultural green water use (i.e., the rain contribution to crops and grazing) at 3700 km$^3$ yr$^{-1}$ (Hoekstra and Chapagain 2007), 5300 km$^3$/yr (Döll 2009), 15,400 km$^3$/yr (Rost et al. 2008), and 25,000 km$^3$ yr$^{-1}$ (Falkenmark and Rockstrom 2004). Using Falkenmark and Rockstrom's estimate, humans are using in total (green + blue) approximately 29,000 out of the total resource of 113,000 km$^3$ yr$^{-1}$, or 26%, rather larger than the 9.7% use of blue water.

In addition, the remaining green water, although not used directly by people, does provide important ecosystem services to humanity. These include, among others, timber and nontimber products from forests; the maintenance of biological diversity in terrestrial ecosystems; the regulation of gases and temperature by forests, wetlands, and grasslands; and the recreational, cultural, and aesthetic benefits from intact terrestrial ecosystems. All of these are dependent on green water flows that allow plants to grow.

## Water Use at the Country Scale

Country-level data on water withdrawals (but not consumption or reservoir evaporation) are available from AQUASTAT. We can use the AQUASTAT data to try to understand how water use relates to a country's level of development, by plotting per capita domestic, industrial, and agricultural water use against per capita GDP (Figure 3.9). This is important for addressing questions of how water use is likely to change in the future, especially in developing countries.

In examining Figure 3.9, one is struck first, perhaps, by the large range in per capita water withdrawals among different countries. This variability is not a simple function of level of development: both low- and high-income countries show quite a range in withdrawals. It is important to note, however, that while not all poor countries have low withdrawals, the countries with the lowest domestic and industrial withdrawals are all poor. Thus, there are 50 countries with domestic withdrawals under 24 m$^3$ person$^{-1}$ yr$^{-1}$; all 50 of these countries have GDPs less than $8000. Similarly, of the 48 countries with industrial withdrawals under 7 m$^3$ person$^{-1}$ yr$^{-1}$, all but 2 (Cyprus and Malta) have GDPs less than $8000.

The agricultural water use data show less of a pattern in relation to income (Figure 3.9). This is not surprising, given the tremendous range among countries in the factors that control agricultural blue water use,

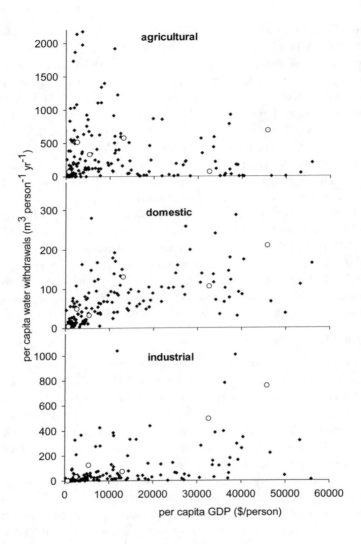

*Figure 3.9.* Country-level per capita water withdrawals for agricultural, domestic, and industrial use, plotted against per capita GDP. Note differences in the y-axis scale between the three parts. Qatar, with a GDP of $87,600, is not shown. Turkmenistan, with a per capita agricultural water use of 5200 m³ person⁻¹ yr⁻¹, is not shown in the top graph. GDP data (Purchasing Power Parity) are from the CIA World Factbook (www.cia.gov/library/publications/the-world-factbook/) and mostly represent the year 2007; water data are from AQUASTAT (2009) and mostly represent the year 2000. Open circles highlight six countries (from left to right): Ethiopia (GDP = $700), India ($2600), China ($5400), Argentina ($13,100), France ($32,600), and the United States ($45,800).

such as the suitability of rain-fed agriculture and the degree to which a country imports or exports grains. Neither agriculture nor the other types of water use show signs of the Environmental Kuznets Curve (Box 3.3).

## Water Footprints

The agricultural (and industrial) water use data shown in Figure 3.9 are a bit misleading, since much of the water used in a country may be used to grow food (or make products) consumed by other countries. A different way to evaluate water use at the country scale is to calculate "water footprints," defined as the total water used directly *and indirectly* by the population. For the domestic sector, this is identical to the water use discussed above, but for agricultural and industrial water use, water footprints are different from simple water use in two ways:

- Instead of calculating the water used by the agricultural and industrial sectors within the country, we now calculate the total water used, both within the country and abroad, to produce the food and products *consumed* within the country. Some countries—those that export more "virtual water" than they import—will have water footprints that are lower than their actual water use, while other countries will have footprints that are higher than their withdrawals.
- In calculating water footprints, we account for agricultural green water as well as blue water use. The water footprint associated with food consumption is calculated as the total water used to produce that food, regardless of whether the food was grown using rain-fed or irrigated agriculture.

The Water Footprint Network, initiated by Arjen Hoekstra at the University of Twente in the Netherlands, has calculated the water footprints of different products (Table 3.3), as well as the per capita water footprints of different countries (Figures 3.10 and 3.11). The factors that appear to affect the size of a country's footprint include the following (Hoekstra and Chapagain 2007):

- total consumption of industrial products and food
- the type of products consumed (especially the amount of meat consumption, which is relatively water-intensive; see Chapter 10)
- the efficiency of agricultural and industrial production (in the location where the country's food and products are produced)

As can be seen in Figure 3.11, industrial water footprints have a strong relationship with GDP, with richer countries consuming more virtual

## Box 3.3. The Environmental Kuznets Curve for Water Use

The Environmental Kuznets Curve (EKC) refers to an inverted-U-shaped relationship between development and environmental degradation, associated with societal changes during the process of industrialization. According to this theory, societies in the first phase of industrialization pay little attention to issues of environmental quality, leading to an increase in pollution. At a later stage, when society is wealthier and pollution impacts are more severe, a transition takes place to a second phase, characterized by increased environmental protection and a decrease in pollution. The transition between these phases (the peak of the inverted U) is referred to as the *turning point*. Hastening and lowering the turning point through technology transfer can potentially allow developing countries to reduce their environmental impact.

While there is good evidence for the existence of the EKC for water and air pollution—where technologies to reduce pollution are available to countries that can afford them—the concept may be less relevant for water use (and resource use more generally). Nonetheless, one might expect that if efficient technologies can allow rich countries to do more with less water, this increased efficiency might show up as an EKC.

The graphs presented in Figure 3.8 do not exhibit an EKC, although this is not surprising given that they are comparing a wide variety of countries at one point in time, rather than looking at countries with similar biophysical conditions over the course of several decades. Only a few studies have looked more rigorously for the EKC in water use. Rock (1998) found that an EKC effect was one of the factors predicting 1990 water use in a set of 68 countries. Bhattarai (2004) found an EKC for irrigated area (not agricultural water use) in Asian countries. Jia et al. (2006) found evidence for an EKC for industrial water use in OECD countries, but the EKC was only present within each country, not across countries, and the turning-point levels of GDP and water use varied dramatically. There are two confounding factors that may make it hard to detect an EKC: the transfer of water-intensive activities (e.g., agriculture) between countries, and the effects of time (efficiency tends to increase over time regardless of level of development). The first factor is accounted for in Figure 3.10, which also does not show an EKC.

As we discuss in subsequent chapters, there is no doubt that technologically advanced countries can reduce their water use through increased efficiency in all sectors. However, there are many other factors that affect country water use. It may be more productive for researchers to directly examine the opportunities for increased efficiency rather than try to prove the existence of the Kuznets Curve for water use.

TABLE 3.3
Average water footprints of different products

Data from Hoekstra and Chapagain (2007). Water footprints of agricultural products are discussed further in Chapter 10.

| Product | Average Water Footprint (liters) |
|---|---|
| tea (1 cup) | 35 |
| beer (1 cup) | 75 |
| wine (1 cup) | 240 |
| apple juice (1 cup) | 240 |
| milk (1 cup) | 250 |
| coffee (1 cup) | 280 |
| potato | 25 |
| bread (1 slice) | 40 |
| apple | 70 |
| hamburger | 2400 |
| paper (1 sheet) | 10 |
| cotton T-shirt | 2000 |
| leather shoes (1 pair) | 8000 |

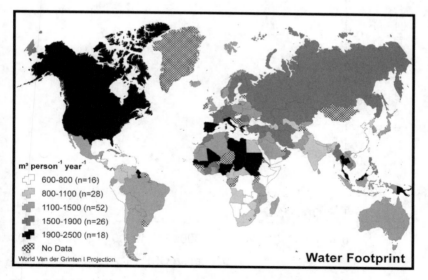

Figure 3.10. Map of water footprint by country. Data source: Chapagain and Hoekstra (2004). Map produced by Stacey Maples, Yale University.

water embodied in products. In contrast, agricultural water footprints have no clear relationship with GDP, but instead appear to be governed primarily by factors such as climate and agricultural practices.

## Water Use in the United States

In this section, we look at water use in the United States since the 1950s. The US is fortunate to have a relatively strong water use data collection program, the National Water-Use Information Program of the USGS

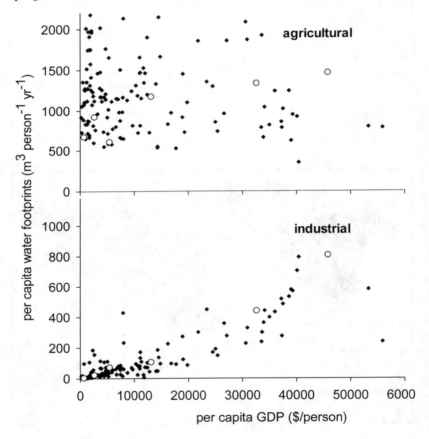

Figure 3.11. Country-level per capita water footprints for consumption of agricultural and industrial products, plotted against per capita GDP. Y-axis scales are the same as in the corresponding parts of Figure 3.9. Qatar, with a GDP of $87,600, is not shown. Data source: Chapagain and Hoekstra (2004). Open circles highlight six countries (from left to right): Ethiopia (GDP = $700), India ($2600), China ($5400), Argentina ($13,100), France ($32,600), and the United States ($45,800).

(water.usgs.gov/watuse/). Every five years since 1950, the USGS has collected and compiled data from the states on various aspects of water use. Unfortunately, the program has been scaled back in recent years, and the reports for the years 2000 and 2005 had several data gaps relative to previous reports, most prominently the lack of data on hydropower (instream) water use, the absence of consumptive use estimates, the elimination of the "commercial use" category, and the reporting of water use only at the state and county scales, not at the watershed scale.

Figure 3.12 illustrates off-stream water use in the US for the year 2005 by category, along with the source of that water (fresh surface water, fresh groundwater, or salt water). Some of the most important features of US water use include the following:

- The biggest water withdrawal in 2005 was for cooling of thermoelectric power plants. This water is generally drawn from surface water and is returned, often to the same river, with only a small loss to evaporation (as well as an increase in temperature). The latest available data on consumptive use (1995) show a 98% return flow for thermoelectric water use.
- Irrigation uses a mixture of surface water and groundwater, with the relative importance of groundwater increasing slightly between 1985 (33%) and 2005 (42%). Irrigation has a low return flow of ~20%, with perhaps 60% of the withdrawn water being consumed through ET on fields and another ~20% being lost during conveyance (1995 data).[11]
- Water delivered by water utilities to residences, institutions, and businesses ("public supply") is mostly (~2/3) surface water and has a fairly high return flow of about 80% (1995 data). In addition, a small portion (~10%) of residential water use is self-supplied from wells ("domestic self-supply").
- In-stream water use for hydropower is not included in Figure 3.12, but the most recent data (1995) suggest that, at ~4400 km$^3$/yr, this use is about 10 times higher than total off-stream use. This number represents about 2.6 times the total runoff in the conterminous US, implying that, on average, each drop of water that reaches the ocean has passed through hydropower turbines two to three times.

---

11. In theory, some of the conveyance loss is consumptive use (evaporation), while some of it could be considered return flow (infiltration).

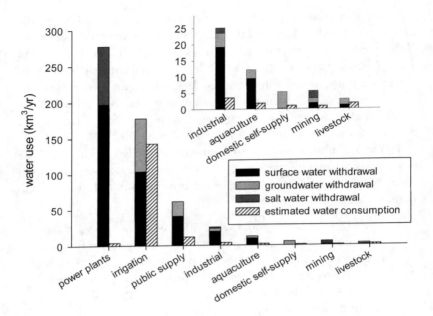

Figure 3.12. US water use by category and water source, 2005. Inset shows the lowest five categories on an expanded scale. "Public supply" includes all water delivered by utilities, which is mostly for domestic use (residential/commercial/institutional) but also includes a small amount of industrial use. (The most recent data available [1995] suggest that industry uses about 12% of the public supply.) "Industry" includes only self-supplied industrial withdrawals. Water consumption is estimated by the author based on 1995 data on percent consumption by sector. Irrigation consumption includes both on-field consumption and conveyance losses. Data sources: Kenny et al. 2009 (water withdrawals), Solley et al. 1998 (percent consumption).

The USGS data can also be used to examine trends over time in US water use (Figure 3.13). The largest change over the last 50 years has been in the use of cooling water in power plants, which rose dramatically through 1975 and then began to decline as more water-efficient cooling systems (closed-loop cooling and air cooling) came online. Other (nonpower plant) industrial water use has also declined over the last 20 to 30 years. Two factors are thought to be responsible: a decline in some water-intensive industries in the US and the requirements of the Clean Water Act, which made discharging wastewater more difficult and expensive and thus encouraged improvements in efficiency. We will return to industrial water use in Chapter 11.

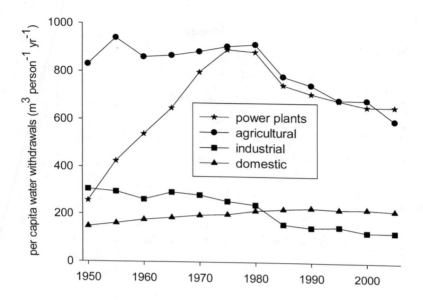

Figure 3.13. Per capita freshwater withdrawals in the US over time. "Agricultural" includes irrigation and livestock; "industrial" includes self-supplied withdrawals by the industrial, mining, commercial, and aquaculture sectors; "domestic" includes self-supplied domestic as well as public supply (which includes some commercial and industrial users in addition to households). For agricultural and domestic uses, per capita withdrawals were calculated from corrected historical withdrawal and population data presented in Kenny et al. (2009). For mining and power plant uses, the Kenny et al. (2009) data include salt water withdrawals; to calculate freshwater withdrawals only, I subtracted the percent salt water use by sector based on historical data (1985–2005) or my estimates (1950–1980; estimate salt water use at 30% of power plant use and 10% of industrial use).

Figure 3.13 also shows that per capita agricultural withdrawals have declined significantly since 1980. This has occurred even as the number of acres irrigated has remained constant or increased slightly. The decline in water use can be attributed to an increase in efficiency, spurred on in part by dropping water tables and increasing energy costs. We will return to agricultural efficiency issues in Chapter 10.

Domestic use is the one sector that has climbed steadily since the 1950s (in per capita terms), as people have moved to larger houses with more water-using appliances and larger lawns. We will return to household efficiency issues in Chapter 9.

Overall, US water withdrawal per capita has declined by about 30%

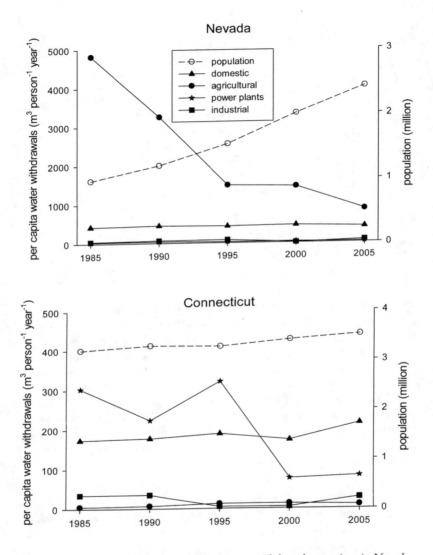

Figure 3.14. Population and per capita freshwater withdrawals over time in Nevada (top) and Connecticut (bottom). Water use categories are as in Figure 3.13. Data source: water.usgs.gov/watuse/.

from its peak in 1980. This has allowed a drop in total water withdrawals despite an increase in population.

We should note, too, that there are large regional differences within the US in the major uses of water and in the trends over time. To demonstrate this, Figure 3.14 presents water use data for a water-scarce western state with rapid population growth (Nevada) and a slowly growing northeastern state with higher water availability (Connecticut). Both states have had relatively minor changes in per capita domestic water use, yet both states have had 30–40% declines in *total* water used, even as their populations have increased. Connecticut has achieved this by reducing power plant water withdrawals, while Nevada has dramatically reduced agricultural water use, in part due to increased efficiency and in part due to an economy that is shifting away from farming and livestock.

## 3.3 Conclusion

In this chapter, we have begun looking at water as a resource—its availability in different parts of the world and the ways in which it is used by people. You should now have an appreciation of the complexities of defining water availability and use, as well as some sense of the geographic distribution of water supply and demand. In addition, a historical perspective is key to understanding the current water crisis, and thus our discussion of the evolution of water use over time (both globally and in the US) is another important part of this chapter.

Even at this point in the book, it has already become clear that the human demand for water is not fixed, but rather varies tremendously in different places and times. On the one hand, this is bad news: global water use increased sixfold during the twentieth century, a problematic pattern that we simply can't afford to repeat during the twenty-first. On the other hand, the range in water use among industrialized countries demonstrates that there are ways to live well with less water; similarly, we can be encouraged by the fact that water use in the US has started to decline, without any drop in our standard of living.

We now turn to two chapters that address more directly two of the main problems associated with water—having too much (floods) and not having enough (scarcity).

# 4

## Water, Water, Everywhere: Dealing with Floods

Although water scarcity problems tend to dominate water management (and this book), *too much water* can also be a problem. In this short chapter, we discuss various aspects of the flooding problem. We begin with definitional issues, then look at trends in flooding impacts, and finally discuss flood management and ways to reduce our vulnerability to floods. The discussion of floods will continue in Chapter 6 in the context of land use and climate change, and in Chapter 8 in the context of the "urban stream syndrome."

### 4.1. Defining Floods

Floods are unusually high water levels that cause the inundation of what we normally think of as dry land. Floods can be classified into three types:

- River floods: When the flow of water in a river is larger than the capacity of the channel to convey it, the river will overflow its banks and inundate its floodplain.
- Urban floods: Stormwater flow in urban areas is often highly modified by land use and by a network of storm sewers meant to convey rainwater to streams. During rainfall events, flooding can occur along urban streams or streets or in people's basements. Backups of storm or sanitary sewer lines are also common.

- Coastal floods: Coastal storm surges can lead to flooding with sea-water, driven primarily by high winds in a landward direction, often combined with high tides. This was the case, for example, with the flooding of New Orleans by Hurricane Katrina in 2005.

Precipitation is obviously a key factor in noncoastal flood events. High-intensity, short-duration events extending over small areas (thunderstorms) can result in rapid stormflow and can cause short-term flooding in small streams, especially in urban areas. In contrast, flooding in large rivers typically lasts longer and is caused by rainfall of longer duration and greater spatial extent. In addition, rain-on-snow events can increase the amount of water flowing to streams by adding large volumes of melted snow.

The most commonly used tool for describing the likelihood of floods of different magnitudes is *flood frequency analysis*. This type of hydrologic analysis gives us terms such as the *100-year flood*. It involves using existing flow data for the system of interest to deduce the likelihoods of different events.

Flood frequency analysis for a given river begins by calculating the highest flow for each year of record, preferably using instantaneous or 15-minute average flows, rather than daily flows. These annual peak flows are ranked from highest to lowest and assigned a probability based on this rank, with the highest (most unusual) flows being assigned a low probability of being exceeded and the lowest floods being assigned a high probability.[1] These are plotted, as shown in Figure 4.1, and a curve is fitted to the points, allowing extrapolation.

A flood's *recurrence interval*, T, is defined as $1/p$, or the inverse of the annual probability. In other words, a flood with an annual probability of 1% is defined as the 100-year flood. To the public, though, the phrase 100-year flood may convey a false sense of regularity, and some managers and scientists have argued that the term should not be used. In any case, it is important to understand that flood events are assumed to occur randomly and to follow the laws of probability (see Box 4.1). An annual probability (or recurrence interval) needs to be presented in that context.

There are two fundamental weaknesses in the flood frequency analysis approach:

- The available hydrologic data often cover a time period that is short relative to the flood frequencies of interest. The guidelines

---

1. The probability assigned is $m/(n+1)$, where m is the rank and n is the number of observations.

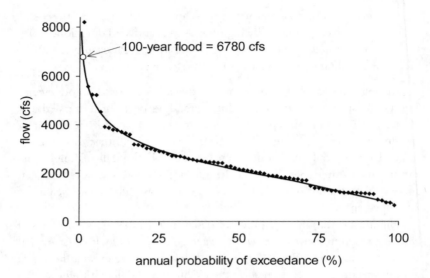

*Figure 4.1.* Flood frequency analysis for the Quinnipiac River. Position on the *x*-axis gives the annual probability that the peak flood will exceed the given flow value (*y*-axis). Points indicate annual instantaneous peak flows, water years 1931–2005 (data source: USGS 2009, station 01196500). Curve is log Pearson III fit to the data. Open circle indicates calculated 100-year flood (1% annual probability).

for flood analysis in the US recommend a minimum of 50 years of record for calculating 100-year floods, but recognize that available records are often considerably shorter. Even when >50 years of data are available, the random nature of flood events can lead to large errors in the calculated 100-year flood.

- The method implicitly assumes that the past is a good guide to the future. In an age of rapid change—both global climate change and local land use change—this assumption is probably not valid for most rivers, as we discuss in Chapter 6. The 100-year flood is a moving target, which means we have no good way to calculate it.

Merz and Bloschl (2008) have suggested replacing flood frequency analysis with *flood frequency hydrology*, which would focus on expanding the methodology from a strictly statistical one to an approach that takes advantage of other types of information, including both qualitative records of floods and an understanding of the watershed processes that cause floods.

## Box 4.1. Probability and Flooding

A few exercises may help illustrate the use of probability principles with flood frequency analysis.

**1. The 2-year flood for the Quinnipiac River is 2140 cfs. What is the probability of getting a peak flow of at least 2140 cfs this year?**

A 2-year flood is the flood with an annual probability of 1/2 = 50%, so the chances of getting a flow of at least 2140 cfs this year is 50%.

**2. What is the probability of getting a peak flow of at least 2140 cfs (the 2-year flood) in both this year and next year?**

The probability of reaching 2140 cfs in any year is 50%, so the probability of reaching it 2 years in a row is 0.5*0.5 = 25% (just like the chance of having 2 girls in 2 pregnancies).

**3. What is the probability of not reaching a peak flow of at least 2140 cfs (the 2-year flood) over the next 3 years?**

The probability of not reaching 2140 cfs in a year is 1–0.5 = 50%. The probability of not reaching 2140 cfs over 3 years is 0.5*0.5*0.5 = 12.5% (just like the probability of *not* having a single girl in 3 pregnancies).

**4. What is the probability of reaching the 100-year flood level sometime in the next 30 years (lifetime of a mortgage)?**

The annual probability of *not* reaching the 100-year flood is 99%, so the probability of not reaching the 100-year flood 30 years in a row is $(0.99)^{30} = 0.74$. This means that the probability of reaching the 100-year flood sometime during that 30-year period is 1–0.74, or 26%.

**5. There was a 100-year flood last year. What is the probability of reaching the 100-year flood level again this year?**

The events are assumed to be independent and randomly distributed, so last year's flood does not affect this year's probabilities. The probability of reaching the 100-year flood this year remains 1%.

---

One alternative to flood frequency analysis is to estimate the *probable maximum flood* (PMF), defined as the maximum flood that is likely to occur in a given area under the most extreme conditions. This is derived from the *probable maximum precipitation* (PMP), estimated by meteorologists based on the maximum amount of moisture that the atmosphere can hold. The amount of the PMP that will run off to form the PMF is then estimated based on watershed soils, land use, and topography. The PMF is typically used in situations where a conservative estimate of possible flooding is needed, such as in dam safety design.

## 4.2. Flooding Impacts and Trends

Not all floods are damaging. Annual flooding is part of the natural pulse of many river systems and can bring renewal to ecosystems and to the human communities that are dependent on them.

However, flooding can also cause loss of life and extensive property damage. Floods can devastate entire communities by inundating streets, cars, houses, and public buildings, and by causing erosion and landslides. When floods occur suddenly, people can literally be swept off and drowned. With adequate warning, people can often escape to higher ground, but may return after the flood to find their homes and belongings gone or suffering from extensive water and mud damage. Besides the obvious impacts, flooding can also lead to disease and hardship as large numbers of people struggle to find clean water and sanitation, especially if the water supply infrastructure was damaged or contaminated with sewage or toxic chemical spills.

Assessing the extent of these impacts globally over time is difficult, since information is often incomplete and methods of evaluating and reporting impacts often differ from country to country. Three entities that compile data on floods and their impacts are the insurance company Munich Re, which has gathered data since 1950 on a variety of natural hazards, especially the most devastating events; the Dartmouth Flood Observatory, which uses global news reports to compile data on floods from 1980 to the present; and the US National Weather Service, which has compiled flood damage data in the US since 1929.

According to Munich Re,[2] there were 173 flood events between 1950 and 2008 that qualified as "great natural disasters." These events caused approximately 820,000 deaths and $1100 billion in damage. In addition, there are more than 100 floods each year that qualify as "natural catastrophes" (Munich Re 2009). Munich Re also notes a strong trend of increasing damages from disasters of all types.

The Dartmouth Flood Observatory calculates a magnitude for each flood recorded, based on its duration, spatial extent, and severity. As shown in Figure 4.2, there seems to be an increasing trend in the annual

---

2. www.munichre.com/en/ts/geo_risks/natcatservice/long-term_statistics_since_1950/default.aspx

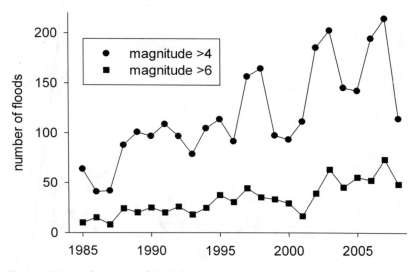

*Figure 4.2*. Annual number of floods having magnitudes greater than 4 or 6. Data source: Dartmouth Flood Observatory (www.dartmouth.edu/~floods/).

numbers of high-magnitude floods, although there is also a great deal of year-to-year variability.

The National Weather Service data on annual flood damages in the US show a similar pattern: high variability from year to year but a clear increase over time from 1929 to 2003 (Figure 4.3).

The apparent increases in flooding have aroused great concern. While methodological issues (improved flood reporting over time, changes in classifications of damage) may contribute to this trend, the consensus among scientists is that the increase is real. However, there is an active dispute over the underlying causes for this trend. Some see it as providing evidence for the role of climate change in causing more severe floods. Others note that the evidence for increased flood *damage* is much stronger than the evidence for increases in physical flooding and attribute the trends noted above primarily to the fact that more people and property are in harm's way. We will postpone our discussion of climate change (and watershed land use change) until Chapter 6 and focus here on the ways that people manage flood risk.

## 4.3. Flood Management

How humans interact with floods and flood-prone areas can have a significant effect on both the physical flood itself and the damage that it

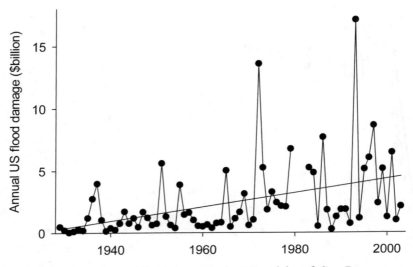

*Figure 4.3*. Estimated annual flood damage in the US, with best-fit line. Damages are expressed in constant 1995 dollars. Data source: Pielke et al. 2002.

does. One key factor is the issue of *vulnerability*—how different settlement patterns and flood management approaches affect the damage that is done by a flood of a given size.

## Vulnerability

It is important to recognize that the damage associated with flooding is a function of both the physical event itself and the ways that humans live relative to that event. Simply put, if a river floods but there are no people or structures in the floodplain, there is no socioeconomic damage. There may still be temporary devastation to natural ecosystems (e.g., damage to trees, erosion of marshes), but these ecosystems tend to be adapted to periodic flooding, and the long-term ecosystem effects tend to be positive.

One way to think about the damage caused by floods (and other natural hazards) is to say that the *risk* associated with flooding is the product of two factors: the *likelihood* of a flood of a given magnitude and the *consequences* of that flood for people. The likelihood is determined by both natural and anthropogenic factors, but the consequence is primarily determined by human behavior. Or, in slightly different terms, *outcome risk* is the product of *event risk* and *vulnerability* (Pielke et al. 2005). It is important to recognize also that vulnerability (or consequence) in turn has two components (Few 2006): physical vulnerability (whether a person will be exposed to a given flood) and social vulnerability (the ability of the individual or community to adapt to the flood, either ex ante or ex post).[3]

## Flood Management Over Time

Humans have always been drawn to floodplains and riverbanks—for water withdrawals, navigation, productive soils, and spiritual sustenance. For most of human history, people have generally used flood-prone areas in ways that were compatible with periodic flooding, such as planting crops, rather than building houses. In fact, in areas where seasonal river flooding is a regular occurrence (e.g., Mekong Delta), societies have developed specific adaptations to live with the flood and take advantage of its benefits, though these adaptations are not perfect, and floods—especially ones with less frequent recurrence intervals—have always taken their toll.

At the same time, there has always been a human desire to control nature, and some rivers (e.g., Huang He) have a long history of human modification in an attempt to control flooding. It was only recently, however, that this desire to control floods was coupled with technology that seemed capable of doing so on a large scale.

In many developed countries, the twentieth century saw a change in attitude and behavior with regard to floods. Specifically, there emerged a pattern of building in floodplains and coastal areas, and protecting those investments with large public works. The development of floodplains was driven both by population growth and by the belief that technology could in fact provide flood protection. The technology deployed generally involved massive infrastructure projects, including levees along riverbanks, flood-control dams, floodways to transport floodwaters, and modifications to river channels (debris removal, deepening, straightening, and hardening). There was little or no recognition that floodplains are an integral part of the river system and that depriving a river of its floodplain will ultimately backfire, both in terms of flood protection and in terms of ecosystem health.

This structural approach to flood management has led to changes in both the consequences and the likelihood of flooding. The former is fairly obvious: more development in the floodplain means more property (and lives) in harm's way, and thus more damage when flooding does occur; this is probably the single most important reason for the increase in flood damage over time.

---

3. These terms and other risk-related words are used differently by different groups; see Gouldby and Samuels 2005.

The change in flooding likelihood is more complex. While structural flood control has had its successes in reducing the magnitude of flooding, this struggle against the power of water has often been a losing battle; as the saying goes, "Nature bats last." Our actions—depriving rivers of their floodplains, trying to move water more quickly, paving over upstream watersheds—have often led to larger and more devastating floods, especially in downstream communities. In addition, levees (and to a lesser extent dams) have a tendency to fail under extreme conditions, exacerbating the problem by suddenly releasing large amounts of water.

The structural flood control movement has had strong political and psychological momentum. Flood control has encouraged development in flood-prone areas, which in turn has created a strong imperative to invest more and more in flood control. When flood protection has failed, the response has often been to build it higher and stronger, rather than to rethink the effort.

In developing countries, some of the same dynamics have played out, but population growth, rather than technology, has been the main driver. The growth in population has pushed people—usually the poorest sectors of society—into vulnerable locations near rivers and along coastlines (Box 4.2).

## Flood Control in the United States

The geographical, historical, and philosophical epicenter of flood management in the US is the Mississippi, the river that gathers water from half the country and moves it down through Louisiana into the Gulf of Mexico. In the lower part of the watershed, the need to protect people from the river goes back to the early days of New Orleans, the city that represented the first attempt to create a large permanent settlement in this highly flood-prone area. By 1727, a 3-foot levee protecting New Orleans was complete, although it was overtopped periodically during large floods.

Over the next 200 years, levees held sway as the technology of choice for protecting land in the lower Mississippi. As agriculture and commerce spread, continuous levees eventually covered both banks of the river. In addition, distributaries that would naturally carry some of the floodwaters were blocked off in order to fully confine the river to its channel. This approach failed again and again when large floods broke through levees at their weakest spot, leading to an "arms race" to build levees higher and higher, until they reached 20 feet in many places.

## Box 4.2. Flooding, Vulnerability, and Adaptation in Bangladesh

Bangladesh, sitting at the confluence and delta of the Ganges, Brahmaputra, and Meghna rivers, is one of the most flood-prone countries in the world. Seasonal river flooding occurs almost every year in the Bangladeshi portions of these rivers as a result of monsoon rains in Bangladesh and upstream in India and Nepal. Of the three rivers, the Brahmaputra typically floods highest and first (June–August), but the contribution from the Ganges can also be very large and usually overlaps, at least partially, with the Brahmaputra flood period. It is estimated (Mirza 2002) that 21% of the area of Bangladesh is flooded in a "normal" flood year (50% annual probability = 2-year recurrence interval), while more than 34% of the country is flooded in a "catastrophic" flood year (5% annual probability). Climate models suggest that catastrophic flooding is likely to become more frequent as a result of global warming (Mirza 2002).

Normal floods are thought to have net positive effects on the country, due to adaptations designed to take advantage of floodwaters for agriculture and fishing. However, more severe floods, such as those in 1988 and 1998, can cause extensive damage to crops, structures, and people.

Agricultural adaptations include a complex system of cropping, in which different rice and wheat varieties are planted at different times and elevations. In response to years of higher-than-average flooding, farmers tend to shift from high-yielding rice varieties to local, more flood-tolerant varieties and also tend to increase their postflood planting in order to take advantage of greater soil moisture and compensate for lower flood-season yields (Uchida and Ando 2007). Farmers (and others) would benefit from better forecasts of flood timing and magnitude, as current flood forecasts have lead times of only 3 days (Chowdhury and Ward 2007). Structural adaptations to flooding include earthen levees and flood-control, drainage, and irrigation (FCDI) schemes that use sluice gates to exert some control over water levels in fields and ponds. Farmers and fishers prefer different operating protocols for the FCDI schemes, which can lead to conflict (Halls et al. 2008).

Lower socioeconomic groups tend to suffer more from flood damage. They experience higher inundation levels, suffer more flood damage as a share of household income, and take fewer preventative measures (Brouwer et al. 2007). They are also more likely to suffer from flood-related diarrhea due to contamination of drinking water sources (Hashizume et al. 2008).

As the levees grew higher, the need to integrate different efforts along the river became apparent. Control over levee construction and maintenance shifted over time from individual landowners to levee districts to the US Army Corps of Engineers, which in 1879 was given overall control of this war against the river.

The devastating Mississippi River floods of 1927 led Congress to pass the Flood Control Act of 1928, which expanded the involvement of the federal government (and federal dollars) in flood control nationwide and emphasized four "hard path" technologies: levees; dams, especially in large tributaries, to hold back floodwaters; channel "improvements" such as meander cutoffs to allow water to move more quickly downstream (and improve navigation); and floodways to which water could be directed from the main channel during large floods.

To this day, flood control in the Mississippi, and in the US more broadly, still relies heavily on those same tools, although they have been supplemented with some attempts to plan development in ways that reduce both downstream flooding and vulnerability. The main tool for this latter approach was the 1968 Flood Insurance Act, which was passed as a response to the increasing amounts of federal money that were being spent in disaster relief after floods. The act created the National Flood Insurance Program (NFIP), administered by the Federal Emergency Management Agency (FEMA), which

- produces Flood Insurance Rate Maps (FIRMs), which show the extent of the 100-year floodplain, that is, the area that would be flooded in a 100-year flood;
- provides subsidized flood insurance to structures within the 100-year floodplain that were built before the maps were created ("pre-FIRM" structures)—under the condition that the local community "joins" the NFIP and develops floodplain management regulations for new buildings; and
- provides nonsubsidized flood insurance to new buildings ("post-FIRM" structures) in communities that have joined the NFIP.

In 1973, additional legislation was passed in response to the low level of participation in the NFIP. The Flood Disaster Protection Act required flood insurance as a condition for mortgages in the 100-year floodplain in communities that have joined the NFIP. It also prohibited certain kinds of federal funding to floodplain communities that have not joined the NFIP. This legislation dramatically increased the number of communities and structures participating in the NFIP.

Overall, the NFIP has had some successes, including an estimated $1 billion annually in reduced flood damages (AIR/NFIP 2006). Still, the program has in some ways had the opposite of its intended effect, as indicated by some disturbing observations. First, even 40 years after its inception, about 25% of the structures in the NFIP are pre-FIRM structures that do not necessarily meet community floodplain regulations but are being encouraged to stay in the floodplain through federal subsidies. In addition, the relatively large number of "repetitive loss structures"— buildings that have had multiple claims on the NFIP—implies that we are still rebuilding structures in places where we shouldn't be. Third, when flooding happens to people who have chosen not to buy flood insurance, the federal government still usually steps in to provide disaster assistance, despite the perverse incentives that this provides. Finally, an accurate assessment of flood risk is a backbone of the program, but updating FIRMs to reflect changing conditions is an expensive effort that is lagging behind where it should be.

One assumption behind the NFIP is that local control of floodplain development is effective. Some 20,000 communities are part of the NFIP and have each developed their own floodplain regulations, though they must conform to the basic requirements of the NFIP (including, for example, a ban on development in the "floodway," defined as the area that can convey the 100-year flood without an increase in water level of more than 1 foot). While FEMA attempts to ensure community compliance, it has limited resources to do so.

The devastating 1993 Midwest floods in the Upper Mississippi Basin caused about $30 billion in damage and led to a further reevaluation of the US approach to flood management. The Interagency Floodplain Management Review Committee (known as the Galloway Commission, after its executive director) called for an increased emphasis on floodplain preservation and vulnerability reduction. But agencies are slow to change, and in 2005 Gerald Galloway noted that the changes at the Army Corps of Engineers had been "evolutionary, not revolutionary" (Galloway 2005).

## The Future of Flood Management

What can be done to better manage floods? There has been a shift in many countries recently from a philosophy of "flood defense" to "flood risk management" (Samuels et al. 2006, Wheater 2006). This involves recognizing that even our best, most expensive efforts to control rivers will not provide us with complete protection against the power of nature, especially in an age of changing climate. Indeed, the false assurance that these efforts provide is part of the problem. The philosophy shift also involves understanding that river flooding provides benefits to ecosystems, not just risks to people (see Chapter 8). It involves focusing less on building large levees and dams, and more on nonstructural measures, and, in particular, dealing with land use and vulnerability in a more thoughtful manner.

Perhaps the greatest benefit from the new philosophy comes from reassessing our patterns of settlement and how they make us more or less vulnerable to the effects of floods. The principles underlying this reassessment are easy to articulate, but hard to implement:

- Land use decisions need to take into account the best available information on flood risks at a given site, including the potential for future changes in flooding as a result of climate change. At the same time, it is critical to recognize that our estimates of flood risk come with a certain amount of irreducible uncertainty.
- Government needs to provide some level of flood protection to existing communities, but this should not be understood as a blanket assurance of complete protection. Government has no obligation to provide flood protection to new structures, as long as it is providing reasonably accurate and updated information on flood risks.
- Ways must be found of "living with water," that is, of accepting that certain areas will be periodically flooded and that structures and communities in those areas must adapt to that fact. The best land uses for floodplains and other flood-prone areas are those that are compatible with periodic flooding and that take advantage of their natural productivity.

## 4.4. Conclusion

This chapter has provided an introduction to the flooding problem, focusing on the issues of vulnerability and flood management. The increase in flood damage over time is complex and has many causes, but increased vulnerability and poor management are certainly among them. Biophysical factors are also important; these are inextricably tied to larger issues, particularly climate change and land use change, and will be dealt with in Chapters 6 and 8. Still, it should already be clear from this chapter that there is much that we can do, in both developing and developed countries, to better adapt to flooding and manage the risks that it poses. We now turn to the opposite problem—that of insufficient water.

# 5

## Nor Any Drop to Drink: Water Scarcity

Given the fundamental need for water in all aspects of human society, the lack of sufficient water to meet these needs poses a significant threat. In this chapter, we define scarcity more precisely and then assess the extent to which scarcity is currently a problem in different locations. We also deal with two specific aspects of the scarcity problem: overuse of groundwater and periodic droughts.

### 5.1. Defining Scarcity

Water scarcity has multiple aspects, and different definitions of scarcity have tried to capture one or more of these facets. In this section, we explore the Falkenmark and the WTA indicators of water scarcity. Both deal with blue water (as opposed to green water) and can be calculated at a variety of scales.

### Falkenmark Indicator

Falkenmark (1986) originally introduced a *water crowding* indicator, defined as the number of people sharing a unit of water (people per annual MCM of water). She argued that as this indicator increased, there would be an increase in societal water stresses, such as water pollution and conflict. This indicator was later inverted from *people per unit water* to *water availability per capita*, with threshold levels (inverted from the original values) shown in Table 5.1. In its current state, the Falkenmark indicator draws our attention to one of the most basic aspects of water scarcity:

how much water is available to meet the needs of each person. The values shown in Table 5.1 are not meant to indicate how much water each person actually needs to use directly, but rather to identify thresholds of availability below which there tend to be problems with obtaining sufficient water for all societal and ecosystem needs.

One of the weaknesses of the Falkenmark indicator is that it does not take into account differences in water needs among different countries or regions. The thresholds in Table 5.1 apply both to countries where conditions are conducive for rain-fed agriculture and to those where all food must be grown using irrigation. These two groups of countries have very different blue water needs, yet the definition of scarcity is tied to the identical level of blue water availability.

## WTA Indicator

The WTA (*withdrawal to availability*) indicator, also referred to as the *UN indicator* or the *criticality ratio*, is perhaps the most widely used indicator of water scarcity. It is defined as the ratio of annual water withdrawals to annual water availability and uses the threshold levels shown in Table 5.1 to define levels of stress. This indicator takes into account the different water needs of different countries by using the water withdrawals themselves as the numerator. The WTA indicator draws our attention to the problems that arise when a country is using a significant fraction of its available water supply: at levels of 40%, we certainly expect to see symptoms such as ecosystem degradation, water conflict, and difficulties in supplying adequate water to all users at all times.

Unlike the Falkenmark indicator, which focuses on natural water endowment, the WTA indicator focuses on how countries are using that endowment. Thus, a country with apparently ample water resources per person by the Falkenmark indicator may be considered stressed by the

TABLE 5.1

Definitions of different levels of water scarcity using the Falkenmark and WTA indicators

The lower levels of water scarcity are also sometimes referred to as "water stress." (TRWR = total renewable water resources; see Chapter 3.)

| Scarcity Level | Falkenmark Indicator ($m^3$ person$^{-1}$ yr$^{-1}$) | WTA Indicator (withdrawal / TRWR) |
|---|---|---|
| none | >1700 | <10% |
| slight | 1000–1700 | 10–20% |
| moderate | 500–1000 | 20–40% |
| severe | <500 | >40% |

WTA indicator if it is using a great deal of water per person, either because it is wasteful or because it is exporting water-intensive products.

The WTA indicator has several weaknesses:

- Data quality: Since the WTA indicator requires data on water use as well as water availability, it is subject to all the problems associated with collecting data on water use.
- Timing: As discussed in Chapter 3, different regions have different relationships between annual and dry season water availability. A country with fairly uniform water availability throughout the year may be able to get away with using 40% of its annual water resource, while a highly seasonal country may find that it is experiencing dry season water problems even when it is using only 20% of the annual resource. This issue can be dealt with by calculating the value of the WTA indicator by season, although the data may not always be available to do so.
- Appropriateness of water use: The WTA indicator takes as a given the current levels of water withdrawals, without giving any indication as to whether these levels are appropriate. Thus, a poor country that has not developed sufficient water infrastructure to meet the needs of its population may appear to be unstressed by the WTA indicator, since water withdrawals are so low—lower than they should be.

## 5.2. Assessing Water Scarcity at Different Scales

Using the indicators defined above, we can assess water scarcity at different spatial scales.

### Global

Calculating water scarcity indicators for the globe as a whole leads to the conclusion that Earth is not water scarce. The Falkenmark indicator, calculated using the world population in 2008 and our best estimate of global runoff (see Chapter 3), is 42,000 $km^3$ $yr^{-1}$/6.7 billion people = 6300 $m^3$ $person^{-1}$ $yr^{-1}$. The WTA indicator, calculated using the most recent water use data,[1] is 4077 $km^3$ $yr^{-1}$/42,000 $km^3$ $yr^{-1}$ = 9.7%. Both of these are in the "no scarcity" category (Table 5.1).

---

1. The estimate of 4077 $km^3$ $yr^{-1}$ total water use comes from adding up the most recent FAO country-level data (see section 3.2), which mostly represent the year 2000, and adding in Shiklomanov's estimate of reservoir evaporation for the year 2000.

## Country Scale

At the country scale (Figures 5.1 and 5.2), we find that most countries are in the "no scarcity" category of the Falkenmark indicator, but a total of 45 countries have some level of water stress, with 19 in the severe scarcity category. The highest-scarcity countries cover a range of income levels (as represented by per capita GDP), but all are Middle Eastern, North African, or small island countries with low levels of natural water resources.

Evaluating the WTA indicator at the country scale (Figures 5.3 and 5.4) shows that 67 countries have some level of water stress, with 30 in the severe scarcity category (withdrawals more than 40% of availability). In fact, a number of countries (14) are withdrawing more than 100% of their TRWR, either by using groundwater unsustainably (e.g., Uzbekistan, Libya) or by desalination (e.g., Malta) or both (e.g., Kuwait).

It is interesting that more countries are considered water-scarce by the WTA indicator than by the Falkenmark indicator. If we accept the validity of the threshold levels shown in Table 5.1, this may indicate that

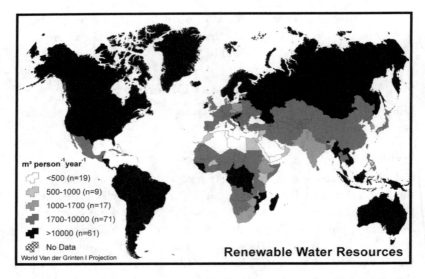

*Figure 5.1.* Map of per capita water availability (Falkenmark water scarcity indicator) by country. Levels of scarcity: severe (<500), moderate (500–1000), slight (1000–1700), none (>1700). The number of countries in each category is shown in the legend. Data source: AQUASTAT 2009. Map produced by Stacey Maples, Yale University.

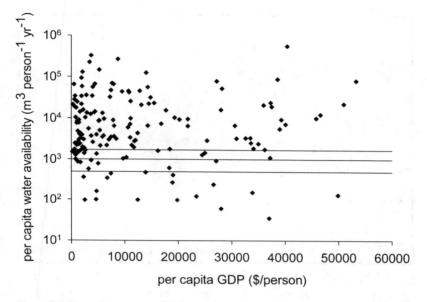

*Figure 5.2.* Per capita water availability at the country scale. The three lines indicate the thresholds for the Falkenmark water scarcity indicator, with the most severe scarcity in the bottom of the figure. Note log scale. Four countries are not shown: Kuwait (off-scale to the bottom), Greenland (off-scale to the top), Luxembourg and Qatar (off-scale to the right). Data source: AQUASTAT 2009.

many countries have a reasonable level of water availability (Falkenmark), but are using too much water (WTA). The United States is an interesting example, with abundant water resources ($>10,000$ m$^3$ person$^{-1}$ yr$^{-1}$) but a WTA ratio (16%) that puts it in the slightly stressed category.

## Grid Scale

Two fine-scale global models of water availability and use have been developed: the Water Balance Model of the Water Systems Analysis Group at the University of New Hampshire (UNH); and the WaterGAP2 model of the University of Kassel, Germany. Each of these divides the earth into a grid, with cells that are 0.5° on a side,[2] and uses globally available data to calculate water availability and withdrawals within each grid cell. Water availability consists of runoff generated within the cell as well as runoff

---

2. 0.5° is about 55 km at the equator, decreasing as you move to higher latitudes.

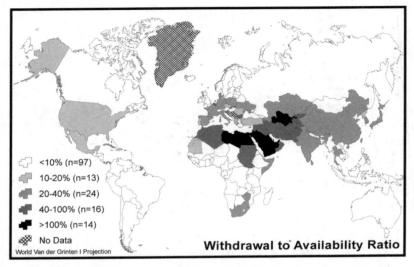

*Figure 5.3.* Map of withdrawal-to-availability ratio (WTA water scarcity indicator) by country. Levels of scarcity: none (<10%), slight (10–20%), moderate (20–40%), and severe (>40%). The number of countries in each category is shown in the legend. Data source: AQUASTAT 2009. Map produced by Stacey Maples, Yale University.

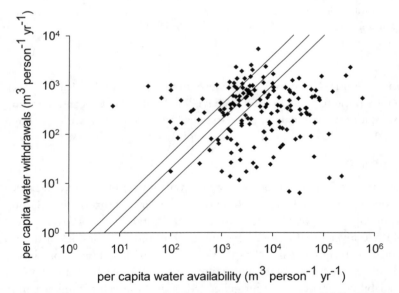

*Figure 5.4.* Country-level water withdrawals plotted against water availability. The three lines indicate the thresholds for the WTA water scarcity indicator, with the most severe scarcity in the upper left of the figure. Note log scales. Data source: AQUASTAT 2009.

that flows into the cell from upstream, with water moving downstream in the models along digital versions of the world's river networks. Water withdrawals for each cell are calculated by sector and can be estimated for different future scenarios.

Vorosmarty et al. (2000) used the UNH model to identify grid cells that exhibit severe water scarcity according to the WTA indicator, that is, cells where withdrawal is >40% of availability. They estimate that the number of people living in high-scarcity grid cells in 1995 was ~1.8 billion (mostly in northern China, western India, Pakistan, the Middle East, North Africa, and western North America), compared to ~0.45 billion people living in high-scarcity *countries*.

Vorosmarty et al. (2000) argue that the country-level estimates mask a huge amount of variability *within* countries, especially for large countries, and that the grid-cell estimates better reflect the local water situation. However, given that countries can—and do—move water from water-rich to water-scarce areas, it is not clear that 50 × 50 km grid cells are really the right scale for expressing the actual water supply available to a population. Nonetheless, there are environmental and financial costs to water transport, so the grid-scale analyses might be best viewed as identifying areas where these costs are likely to be large.

## River Basin Scale

The river basin is the natural unit for assessing water scarcity, since it defines a discrete shared water resource (Chapter 2). If we want to avoid interbasin transfers and move toward watershed management, then the inhabitants of a basin—whether they are citizens of one country or many—must live within the limits imposed by water availability within the basin.

While water availability data are readily accessible at the basin scale (in the form of river flows), data on water use are typically gathered at the country level, so we must rely on models, rather than raw data, for basin-scale analysis, at least for the WTA indicator.[3]

Alcamo et al. (2003) used the WaterGAP2 model to calculate that, as of 1995, 2.1 billion people were living in river basins with severe water scarcity by the WTA indicator (see Box 5.1). Given that 0.45 billion

---

3. For the Falkenmark indicator (which does not require data on water use), Revenga et al. (2000) used basin-level runoff data together with gridded population data to calculate that ~1.1 billion people live in high-scarcity river basins (Huang, Hai, Huai, Jordan, and several areas without significant surface water drainage), compared to the ~0.2 billion people living in high-scarcity countries.

people are living in water-scarce countries, this implies that many people who are living in water-scarce basins are simultaneously living in countries that are less water scarce. This can be largely attributed to the fact that several large countries (China, US) include both highly populated arid river basins (Huang, Colorado) and more humid areas. Interbasin water transfers are currently an important part of water management in these countries (see section 7.3).

Some of the world's highly utilized river basins have WTA ratios that are much greater than 40%. *Closed basins*—also referred to as *overallocated basins*—are defined as those where all the flow is spoken for so that there is no uncommitted outflow from the basin on an annual basis, corresponding to a WTA of close to 100% (Falkenmark and Molden 2008). (*Closing basins* are those where there is uncommitted outflow in the wet season, but not in the dry season.) Some definitions of basin closure include environmental flows in calculating use, so a basin is defined as closed as soon as human use begins to cut into minimum environmental flows. However, this is difficult to implement given the lack of a scientific and political consensus on the size of the required environmental flows, as we will discuss in section 8.5.

River basins that are commonly understood to be overallocated include the Colorado, Rio Grande, Nile, Limpopo, Jordan, Syr, Amu, Huang, Indus, Cauvery, and Murray-Darling (Alcamo et al. 2003, Comprehensive Assessment 2007). In addition, several areas without much perennial surface water drainage, such as North Africa, also experience WTA ratios that approach or exceed 100%.

## Economic Water Scarcity

Another very important aspect to water scarcity that we have thus far ignored is the lack of access to water at the household level. A large number of people in developing countries do not have access to a reliable supply of safe water for the most basic domestic needs. This is not a result of country-level physical water scarcity: even the countries with the highest water scarcity have enough water to meet residential needs (although they don't have enough water to grow all their own food and allow unconstrained industrial development). Rather, household-level scarcity is a result of shortcomings in economic and institutional capacity: the lack of resources to build and maintain the infrastructure necessary to supply people with clean water.

We will come back to the issue of water (and sanitation) access in Chapter 9, but for now we look at the related phenomenon of *economic*

---

**Box 5.1. Methodological Issues with Global-Scale Models**

It is important to note that Alcamo et al. (2003) break large basins up into different parts, some of which are not true watersheds (see Chapter 2). For example, they identify the lower Nile as water scarce, but not the upper Nile, apparently because cumulative withdrawals only reach 40% of cumulative availability when one gets to the lower part of the basin. One could argue that the Nile as a whole should be considered water scarce, since the WTA indicator is >40% for the basin as a whole. This would imply that Alcamo et al. (2003) are underestimating the true extent of basin-level water scarcity.

Even with Alcamo et al. (2003)'s narrow definition of river basin, their estimate of the population facing basin-level scarcity (2.1 billion) is slightly higher than Vorosmarty et al. (2000)'s estimate of the population facing grid-level scarcity (1.8 billion). This is surprising, given that the basin scale relaxes the grid-scale constraints of assessing water availability only within 50 km. This points to the methodological difficulties and resultant uncertainty associated with global-scale models.

---

*water scarcity*. This term, popularized by the International Water Management Institute (IWMI), refers to situations where economic, rather than physical, constraints are limiting the development of water infrastructure and resulting in a lack of water availability to individual users. Over the years, IWMI has used two different indicators to capture this phenomenon.

The first IWMI indicator of economic water scarcity, calculated at the country scale, was based on how rapidly withdrawals needed to increase in the future (from 1990 to 2025) to satisfy the demands of larger and more developed populations (Seckler et al. 1998). Countries that were not physically water scarce, but needed to increase their withdrawals by more than 100%, were considered to have severe economic scarcity, given the magnitude of the financial and logistical challenge of mobilizing such substantial amounts of new infrastructure. Much of sub-Saharan Africa fell into this category. Countries that needed to increase their withdrawals by 25–100% were classified as experiencing a lesser degree of economic water scarcity; much of Latin America, Africa, Southeast Asia, and Australia fell into this category. Note that this indicator suffers from the inevitable assumptions, complexities, and data problems associated with forecasting future water needs.

The second IWMI indicator of economic water scarcity (Comprehensive Assessment 2007), calculated at the sub-basin scale, is based on the current extent of malnutrition, since malnutrition is linked to inadequate water infrastructure for irrigation. Most of sub-Saharan Africa is again shown to be economically water scarce, along with much smaller parts of Latin America and southern Asia. This indicator is widely cited, but the methodology behind it has not been adequately documented in the published literature.

## 5.3. Groundwater Overdraft

As mentioned above, groundwater is included in a country's renewable water resources (TRWR) only to the extent that it recharges on an annual basis. Yet in many places there are vast stocks of groundwater sitting under our feet, and the temptation to use them has often proven irresistible. When groundwater is used more rapidly than it is being recharged, water tables fall over time, which can lead to a variety of problems:

- Costs (financial and environmental) increase, as wells must be deepened and more energy must be used to pump water from greater depths.
- Irreversible subsidence of the land surface may occur, as pores that were filled with water collapse when the water is pumped out. This subsidence can be patchy and lead to infrastructure problems near the surface, such as buildings tilting or pipes bursting. In coastal areas (e.g., Venice), subsidence can lead to a rise in relative sea level and an increase in tidal flooding.
- A hydraulic gradient is created, which can draw in polluted water from nearby sites or, in coastal areas, draw in seawater to replace the freshwater pumped out (*seawater intrusion*). These contaminants or salts may make the remainder of the aquifer unusable.

Since aquifers typically extend over relatively large areas, there can be a strong equity component to these problems. Those wealthy enough to do so can drill deeper wells and pump out groundwater, drying up the wells of neighbors who don't have the resources to deepen their own wells. Sometimes the poorer neighbors may end up having to buy water pumped from the same deep wells that dried up their own water sources!

And, of course, there is the intergenerational equity problem. Drawing down water tables can't go on forever. Groundwater that is overpumped today is not available for use by the next generation.

Yet there are advantages to groundwater use. It doesn't require the damming of rivers and the building of canals. It is decentralized and locally controlled. It allows farmers to achieve higher yields by always having water available when they need it. It doesn't have evaporation losses associated with storage, like surface water reservoirs do. *Conjunctive use*—in which surface and groundwater are used intelligently together—can lead to greater efficiency and resilience in water systems.

In addition, some have argued that mining the world's fossil groundwater is a necessary and even wise course of action, especially in areas where the reserves are so large that it will take many years to deplete them. The current human needs for water are so pressing that slowly drawing down our groundwater reserves may make sense, if it can buy us time to improve our water management and make the transition to a sustainable future. Foster and Loucks (2006) acknowledge that use of nonrenewable groundwater is physically unsustainable, but suggest that it may be socially sustainable if four criteria are met:

- Groundwater use leads to clear improvements in livelihoods and well-being.
- Short-term benefits outweigh long-term costs.
- Thought has been given to what happens after the resource is depleted.
- Issues of intergenerational equity have been considered.

Implementing these four criteria is a daunting challenge, and most countries that are mining groundwater have not even begun to address them.

Data on groundwater availability and use are hard to come by. Döll and Fiedler (2008) estimate that global groundwater recharge amounts to about 13,000 km³/yr (see Figure 3.1), while Shah et al. (2007) and Döll (2009) estimate that global groundwater withdrawal is currently about 1000–1100 km³/yr. Vrba and van der Gun (2004), from the International Groundwater Resources Assessment Centre (IGRAC), estimate that 14 countries (Algeria, Bahrain, Egypt, Iran, Israel, Libya, Mauritania, Pakistan, Qatar, Saudi Arabia, Tunisia, Turkmenistan, UAE, and Yemen) are currently using more than 100% of their renewable groundwater resource (i.e., the rate of pumping is estimated to exceed the rate of recharge for the country as a whole).[4] Several of these countries rely

---

4. The AQUASTAT database has a slightly different set of 13 countries for which reported groundwater use is greater than the renewable groundwater resource: Bahrain, Djibouti, Egypt, Iran, Jordan, Kuwait, Libya, Oman, Qatar, Saudi Arabia, Tunisia,

heavily on fossil aquifers that have negligible current rates of recharge. For several of the most important fossil aquifers, estimates are available of the total exploitable volume of water in the aquifer (Margat et al. 2006). Calculations based on these estimates show that the time-to-depletion for these aquifers, based on current extraction rates, ranges from 30 years (the Central Kalahari Karroo Sandstone aquifer in Botswana) to >6000 years (the huge Nubian Sandstone formation underlying parts of Egypt, Libya, Sudan, and Chad).

In addition, many other countries are drawing down local or regional nonfossil aquifers, even if they are not using more than 100% of their groundwater in the country-scale analysis. Examples of locations where dropping water tables have been documented include Mexico City, northeastern China, and the Ogallala aquifer in the High Plains of the United States. No comprehensive assessment of the world's aquifers has been carried out.

Recently, a new tool has been developed for remote sensing of changes in groundwater storage, based on local changes in Earth's gravity field recorded by the Gravity Recovery And Climate Experiment (GRACE) satellite, launched in 2002. The first application of GRACE has been to measure groundwater depletion in northern India, where high population density and intensive agriculture have led to dropping water tables. The GRACE data—the first reliable and complete information on the extent of the problem—indicate that over the time period 2002–2008, the area lost about 33 $km^3$ of groundwater per year (Tiwari et al. 2009). Given that annual recharge is estimated at 246 $km^3/yr$, this implies an extraction rate of 279 $km^3/yr$ (113% of recharge), which is considerably higher than previous estimates based on spotty on-the-ground data. Nearby areas of Pakistan and Bangladesh are also part of this hot spot of groundwater depletion.

Groundwater—including fossil groundwater—has an important role to play in future water management. But the current disorderly, unplanned overexploitation of this critical resource is a mistake that we will regret for many years. The race to the bottom has resulted in precious groundwater being used inefficiently and uneconomically. In addition, it has allowed us to temporarily ignore looming water scarcity by patching

---

UAE, and Yemen. However, this list is certainly incomplete, because most countries (including the countries that are on the IGRAC list but not the AQUASTAT-derived list) do not report to AQUASTAT the source of water used (groundwater vs. surface water).

it over with borrowed groundwater, rather than devising sustainable long-term solutions.

## 5.4. Drought

The indicators of scarcity discussed above are generally based on *average* water availability. However, deviations from the average are an extremely important phenomenon and can result in both floods (Chapter 4) and droughts (discussed here).

Droughts can significantly increase the degree of water stress. For example, Vorosmarty et al. (2005) calculate that the number of people in Africa living in severe water scarcity (by the WTA indicator) increases from 174 million under average conditions to 262 million in a 30-year drought.[5]

### Defining Drought

Drought is defined not as low water availability (aridity), but as a deviation from normal water availability. Thus, 500 mm of precipitation in a year could be defined as a drought for one region and as an unusually wet year for another. As it turns out, the areas of the world with lowest rainfall also have the greatest interannual variability and thus the greatest likelihood of experiencing drought (Brown and Lall 2006). But all regions can experience drought, even if it occurs less frequently in some areas and even if its effects differ from region to region.

Droughts are usually classified into four types:

- *Meteorological droughts* are characterized by a deficiency in precipitation relative to normal conditions and are generally the first manifestation of drought.
- In *agricultural droughts*, low soil moisture (caused by meteorological drought) leads to negative impacts on crop growth. Falkenmark and Rockstrom (2004) also use the term *agricultural drought* to refer to low soil moisture resulting from anthropogenic degradation of the soil's moisture-holding ability, due to compaction and loss of topsoil.
- *Hydrological droughts* are characterized by a decrease in surface water flow and groundwater levels due to the cumulative effects of meteorological drought.

---

5. The 30-year drought is the drought with a recurrence interval of 30 years, that is, the drought with an annual probability of 1/30. As we discussed in the context of floods (section 4.1), this should not be taken to mean that this event will occur regularly every 30 years.

- *Socioeconomic drought* refers to the socioeconomic impacts associated with meteorological, agricultural, or hydrological droughts. The most dramatic social and economic impacts generally result from crop failures, which can lead to mass starvation. During the twentieth century, there were seven droughts that led to the deaths of more than half a million people each, giving drought the dubious distinction of being the most devastating of all natural disasters. Other impacts can also be important, including wildfires, a decrease in hydropower production, and strained water supplies for households and industry.

Unlike other natural hazards such as earthquakes or floods, drought occurs gradually, and it can be hard to define exactly when a region enters or exits drought conditions. Several indicators have been developed to quantify the occurrence and magnitude of drought. For example, the *Standardized Precipitation Index* (SPI) quantifies deviations from long-term precipitation records using different time scales (1–12 months) that correspond to impacts on different parts of the hydrologic cycle (e.g., soil moisture, groundwater levels). The *Palmer Drought Severity Index* (PDSI) takes into account both the difference between actual and normal precipitation and the difference between actual and normal temperature (and thus PET). Both the SPI and the PDSI are based on normalized records, so values can be directly compared, even between locations that are very different in their average water availability.

Hydrological drought can be quantified using streamflow statistics—assuming that flow data are available for a relatively long period. Two standard measures of drought are the Q99 and the 7Q10.

The Q99 is the flow that is exceeded 99% of the time, derived from a flow duration curve of the type shown in Figure 3.7. By definition, the river sees flows this low only about 1% of the time, which serves as a good definition of hydrological drought.

The 7Q10 is derived from *low-flow frequency analysis*, which focuses on the annual probability of occurrence of certain low-flow conditions. It is conceptually identical to flood frequency analysis (section 4.1), except that the unit of analysis is commonly the seven-day average flow, since low flows (unlike high flows) typically last for at least a week. To carry out this analysis, the lowest seven-day flow in each year of record is tabulated, and these are ranked from lowest to highest. Each value is then assigned a probability based on its rank, with the most extreme low-flow conditions assigned a low probability and the most common low-flow conditions assigned a high probability.[6] These are then plotted, as shown in Figure 5.5, and a curve is fitted to the points, allowing extrapolation. The most commonly used parameter from this graph is the flow that corresponds to a 10% annual probability of occurrence (or a recurrence interval of 10 years), referred to as the 7Q10.[7] This is often roughly the same value as the Q99.

## Paleoclimatology and Drought Frequency

In order to adequately plan for drought, it is important to understand the frequency and severity of droughts that are likely to occur. Unfortunately, in many regions, adequate precipitation and flow records do not exist.

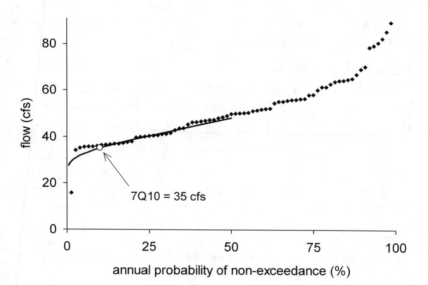

Figure 5.5. Low-flow frequency analysis for the Quinnipiac River. Position on the x-axis gives the annual probability that the 7-day low flow will be lower than the given flow value (y-axis). Points indicate 7-day low flows, based on daily data for water years 1931–2005 (data source: USGS 2009, station 01196500). Curve is log Pearson III fit to the data. Open circle indicates calculated 7Q10 (10% annual probability).

---

6. The probability assigned is m/(n+1), where m is the rank and n is the number of observations.

7. In general, these flow parameters are expressed as dQt, where d indicates the number of days over which flows are averaged and t indicates the recurrence interval. Thus the 14Q20 would be the 14-day low flow with an annual probability of 5%.

Even when these do exist, they are often not long enough to give us a good sense of the frequency of unusual events.

Paleoclimatology, the study of past climates, can be helpful in supplementing the historical record on drought frequency. A variety of proxies can be used to reconstruct past precipitation, temperature, and streamflow, including tree rings, lake sediments, and isotopic tools. Some studies have found that infrequent severe and extended droughts are an important feature of the paleoclimatological record (Box 5.2). Human psychology may interfere with our ability to plan for these types of low-frequency events, since our expectation of the range of conditions that can occur is conditioned by our experience over a limited life span. The challenge of drought preparedness is to get ourselves to take seriously the possibility of an extreme event that we have never experienced and that has a low—but real—probability of occurring.

## Drought Management

We noted in Chapter 4 that flood damage is a function of human behavior as well as the vagaries of nature. The same applies for drought. The risk posed by drought is tied not just to the frequency and severity of rainfall anomalies, but also to the vulnerability and adaptive capacity of human communities. Many factors are thought to affect a society's drought susceptibility, including its degree of dependency on agriculture, its capacity for adaptation, and its capacity in general: financial, human, infrastructural, societal (Alcamo et al. 2008). However, rigorous tests of the importance of these factors have been lacking.

Preparing for drought is an essential task for water resource managers. Some tools that can be used for drought management include the following:

Monitoring and forecasting. Providing early warning of drought conditions can allow water managers and users to take moderate steps early on and avoid the need for drastic steps later. Good weather records, along with a system for analyzing them and publicizing the results, are key to this strategy. Climate models can be useful in understanding the changing likelihood of different levels of drought.

Margin of safety. The *margin of safety* for a given water system is the gap between average water supply and average water demand. The greater the margin of safety, the lower the susceptibility to drought. Conversely, the greater the interannual variability in water supply, the greater the margin

## Box 5.2. Recent Drought in East Africa

As of 2009, Kenya and other East African countries were suffering from a severe drought. Kenya normally experiences two rainy seasons (March–May and October–November). Rainfall has been significantly below average since 2007, and the spring 2009 rains failed completely in much of the country. Rivers and watering holes have gone dry, and trees and grasses have dried up. With little to eat or drink, animals—both domesticated and wild—have died by the thousands. The images of dead elephants have been seen around the world, potentially affecting Kenya's tourism industry. Nomadic pastoralists have lost many of their goats, sheep, and even camels, a devastating blow that has left many in need of food assistance. The national and international response to the emergency has been relatively weak, in part due to Kenya's recent political instability, and there are increasing reports of malnutrition and starvation, especially among children. Oxfam estimates that some 23 million people in Kenya, Ethiopia, Somalia, and Uganda are "facing critical shortages of food and water."

What has caused the Kenyan drought? Three perspectives have been expressed:

- Natural variability: An influential paleoclimatological study (Verschuren et al. 2000) found evidence that over the last 1100 years, Kenya has experienced extended wetter and drier periods. The authors of the study see extended droughts as part of the natural variability of Kenyan climate, and predicted in 2000 that a megadrought was likely sometime in the twenty-first century (Stevens 2000).
- Climate change: Some see the recent drought as part of a climatic shift resulting from anthropogenic emissions of greenhouse gases (Gettleman 2009; see section 6.1).
- Deforestation: Some blame the drought on the destruction in recent years of large areas of Kenya's Mau Forest (Associated Press 2009). In this view, the conversion of forest to croplands has led to a decline in precipitation and to a decrease in the flow of the rivers that originate in the forest and that bring water to downstream ecosystems (see section 6.2).

of safety that is needed. Where margins of safety are inadequate, they can be expanded by either increasing supply or decreasing demand.

Emergency measures. Rather than permanently increasing the margin of safety in order to deal with temporary, occasional scarcity (i.e., droughts), it may make more sense to focus on building in an *implicit margin of safety*, that is, supply- and demand-management tools that can be quickly and temporarily activated in the case of drought. Examples of such tools include the following:

- Pre-agreements with large industrial users that they will temporarily shift to less water-intensive production methods in case of drought
- Cutbacks of water allocations to farmers; these cutbacks need to be made before the planting season to avoid losing crops which have already used some irrigation water; this, in turn, requires good early forecasting of drought
- Development of groundwater supplies that can be tapped in case of drought
- Restrictions on water uses that are permitted under normal conditions, such as residential outdoor water use or flushing of mains by water utilities
- Implementation of higher water prices to encourage conservation

The issue with these approaches is that, in order to be useful for drought management, they must remain emergency measures. If they are implemented routinely—which may make sense for some of these measures—then they are no longer available in case of drought. Thompson (1999), for example, points to the double-edged sword of efficiency improvements by asking, "Are higher-efficiency water-use systems more vulnerable to drought?" His logic is simply that there is less opportunity for emergency water savings in systems where efficiency measures have already been implemented, a phenomenon referred to as *demand hardening*. Cooley et al. (2007), however, argue that conservation measures do not preclude emergency measures during drought (such as behavioral changes and implementation of even better technologies) and can even provide more of a buffer against drought by increasing the normal operating water levels in reservoirs and aquifers.

One advantage that water managers have in the struggle against drought is that any given drought is usually limited geographically. For example, it would be highly improbable for all of the agricultural areas of the US to be suffering from drought in the same year. As a result, average agricultural productivity across the entire country is relatively insensitive to the effects of drought. For smaller countries, this may mean relying more on food imports or food aid during drought years.

## 5.5 Conclusion

In this chapter, we began to explore various aspects of water scarcity, including definitions of scarcity, the geographic extent of current scarcity, groundwater overdraft, and droughts. It should be clear at this point that human populations are distributed very unevenly with respect to available water. Some areas are already suffering from not having sufficient water to meet all human and ecosystem demands. This critical issue will dominate much of the rest of the book, as we examine specific aspects of scarcity and ways to better manage this problem.

Solutions to water-scarcity problems generally fall into three categories:

- Increase supply: This is the traditional response to water scarcity: build new infrastructure to capture, store, transport, and deliver more water. Although these tools still have a place, especially in the least developed countries, we are now much more aware of the shortcomings of large infrastructure projects. Are there other, more sustainable ways to increase supply? We discuss some alternatives in Chapter 7.
- Decrease demand: Conservation and efficiency measures have great potential to solve water-scarcity problems by decreasing the need for water. Conservation tools and how much water they can actually save are discussed by sector (domestic, agricultural, industrial) in Chapters 9, 10, and 11.
- Reallocate water for greater efficiency: *Who gets the water* is an important factor in determining whether we are getting the maximum utility from a limited supply. Reallocation of water to higher valued uses (e.g., from agriculture to cities) can help solve scarcity issues. We discuss water allocation in Chapters 12 and 13.

For now, our discussion of scarcity continues with a look at how scarcity (as well as flooding) is being affected by global and local environmental change.

# 6

## The End of Stationarity: Water in an Era of Global Change

Water is a highly dynamic resource. As discussed in the previous two chapters, humanity has always had to deal with the variability of water: wet seasons and dry seasons, floods and droughts. Yet until recently we have been able to assume that the underlying *patterns* of variability are constant—that we can use the past to plan for the future, that water availability will remain within the range that we know and have planned for.

But now that fundamental assumption of "stationarity" in water supply (Milly et al. 2008) has been called into question by both climate change and land use change. Each of these factors has already had an effect on water quantity and quality, though the effects of climate change to date have sometimes been subtle and hard to prove in the face of natural variability. At the same time, water demand is also changing as a result of rapid population growth. All of these factors must be taken into account in planning for an unknown future. In this chapter, we first address the linked issues of population growth and climate change, and then discuss a variety of different types of land use and their effects on water resources.

### 6.1. Population Growth and Climate Change

#### Population Growth

As discussed in Chapter 3, the rapid growth of population in the twentieth century led to a massive increase in global demand for water. In addition, as society grew wealthier during the twentieth century, we began consuming more water per person, both directly, in the form of greater per capita household water use, and indirectly, in the form of greater consumption of water-intensive foods and other products (though, as noted in section 3.2, the relationship between wealth and water use is a complex one).

As populations and economies continue to grow into the twenty-first century, we need to ask: will these patterns of increasing water use continue, or will they be balanced by increases in efficiency? It is important to note that past projections of water use have generally turned out to be overestimates, because they have projected forward existing consumption patterns without adequately accounting for improvements in efficiency (Gleick 2003). The only thing we can say for certain about the future is that we don't know what will happen! Nonetheless, our forecasting tools have improved over time and it is worth taking a look at current projections, as we do below.

## Climate Change

At the same time, we also need to account for changes in water availability, primarily due to *global climate change*, often referred to as "global warming." Human activities (primarily fossil fuel combustion and tropical deforestation) have led to increasing concentrations of greenhouse gases (primarily $CO_2$) in the atmosphere. This has resulted in a global average temperature rise of ~0.7 °C in the last 100 years and is projected to lead to further temperature increases of somewhere between 1 and 6 °C during this century (IPCC 2007). What effects will this have on water availability? Has climate change already led to alteration of the hydrologic cycle? In tackling these questions, we enter a difficult realm of science, where the nonlinear effects and complex interconnections of the climate system result in high levels of uncertainty.

Higher temperatures are expected to result in higher rates of evapotranspiration and thus more moisture available for precipitation (a "speeding up" of the hydrologic cycle). Climate models suggest that this should lead to an overall increase in precipitation globally by about 1–3% per °C of warming. However, a recent analysis of actual precipitation data for the last 20 years (Wentz et al. 2007) found a much larger increase of ~7% per °C of warming; both the climate models and the precipitation data have their shortcomings (Lambert et al. 2008). It is crucial to note that the past and projected increases in precipitation are not geographically uniform, since spatial patterns of atmospheric circulation are changing, and thus some areas will see (indeed, have already seen) decreases in precipitation.

Predicting the effects of climate change on runoff is more complicated than for precipitation, but is ultimately more important for understanding blue water availability. On the one hand, higher precipitation might lead to greater runoff. On the other hand, higher ET (due to higher temperatures) might lead to a net decrease in runoff.

In addition, several other (nonclimate) anthropogenic activities could lead to changes in runoff (Piao et al. 2007, Gerten et al. 2008). Increased water withdrawals for irrigation could lead to greater water consumption (ET) and thus lower runoff downstream. Deforestation could lead to less opportunity for ET and thus greater runoff (see section 6.2). Higher atmospheric $CO_2$ concentrations might lead to greater transpiration due to increases in plant productivity, but could also depress transpiration through a physiological effect on plant stomates. Of course, each of these effects will vary from location to location.

Has runoff already changed over the last 50–100 years? Incomplete data and the natural year-to-year variability in runoff have made answering this question quite difficult, and disentangling the role of climate change from the other human influences discussed above (consumptive water use, land use change, $CO_2$ effects) adds an additional layer of complication. On the one hand, Labat et al. (2004) found a significant trend of increasing global runoff over the twentieth century, at a rate of roughly 4% per °C of warming. On the other hand, Dai et al. (2009), using a more complete dataset on global river flow, found a slight (not statistically significant) downward trend over the period 1948–2000. Of the world's 200 largest rivers, Dai et al. (2009) found significant downward trends for 45 and significant upward trends for 19. Upward trends dominated in rivers draining to the Arctic, presumably because of decreased snowpack. In general, the changes in runoff seemed to be associated with climatic effects (such as precipitation changes driven by global warming and by El Niño), rather than direct human effects (such as reservoirs and water withdrawals). Similarly, Milliman et al. (2008) analyzed 60 large rivers and found that slightly more than half (34) had changes in runoff that reflected changes in precipitation over the basin (e.g., Mississippi River: 14% increase in precipitation, 31% increase in runoff). However, they identified 17 "deficit" rivers, where increases in anthropogenic withdrawals have led to a disconnect between precipitation and runoff (e.g., Huang He: no significant change in precipitation, 82% decrease in runoff).[1] Döll et al.

---

1. Milliman et al. (2008) also identified 9 "excess" rivers, all high-latitude or high-altitude systems, where runoff increased more than would be expected from precipi-

(2009) modeled the effects of water withdrawals and reservoirs (but not climate), and found a global decrease in runoff of ~1400 km³/year (3.5%), concentrated in certain areas, which are mostly the same as Milliman's deficit rivers.

How will runoff change over the next 50–100 years? General circulation models (GCMs)—the climate models used to understand the complex interactions among the atmosphere, oceans, and land—are not yet very good at accurately predicting the net changes in the hydrologic cycle for particular areas, and different GCMs often give strikingly different results. One approach to dealing with this problem is to use an "ensemble" of different models. Using this approach, Milly et al. (2005) found that there were parts of the world where most of the 12 models in their ensemble agreed on expected changes in runoff over the next 50 years. Specifically, areas that are predicted to experience runoff reductions of 10–30% include the western US (Box 6.1), Central America, southern South America, the Mediterranean area, and southern Africa, while areas that could see 10–40% increases in runoff include the La Plata basin in South America, the Ganges-Brahmaputra-Meghna basin in Asia, eastern equatorial Africa, and high northern latitudes throughout the world. Nohara et al. (2006) carried out a similar analysis with a different ensemble of models and obtained remarkably similar results, which implies that we should take these predictions seriously, despite the inherent uncertainty associated with these types of predictions. Unfortunately, the areas that are predicted to get drier are mostly already water-scarce, while the areas that are expected to get wetter include some areas where flooding is already a significant problem.

One additional climate factor to consider has to do with glaciers and the timing of runoff. Approximately 1 billion people live in areas where snowmelt dominates runoff (Barnett et al. 2005). In these areas, the snowpack (or icepack) serves as a natural reservoir, storing water during the winter months and releasing it in the summer. As temperatures warm, these areas will store less snow. This will shift the seasonal timing of runoff away from the summer months, precisely the time when it is most needed. Unless these basins have adequate anthropogenic storage (e.g., reservoirs), they will be vulnerable to water scarcity. For example,

---

tation changes (e.g., Brahmaputra River: nonsignificant change in precipitation, 25% increase in runoff). They tentatively ascribe this phenomenon to reduced evaporation, perhaps because warming of frozen soil has resulted in more water percolating into the subsurface.

the disappearance of the Chacaltaya glacier in the Bolivian Andes has contributed to the water woes of the cities of El Alto and La Paz (Rosenthal 2009). Note that some snowmelt-dominated basins may experience a short-term *increase* in water availability as glaciers retreat and their meltwaters flow downstream, but in the long term the absence of glaciers and snowpack can be expected to lead to scarcity.

## Forecasting the Combined Effects of Climate and Population

Given that both water withdrawals and water availability are changing, which is more important in controlling future water stress: changes in population/development or climate change? Two research groups have been examining this question, and both agree that in most—but certainly not all—locations, climate change will have less impact than changes in demand.

Vorosmarty et al. (2000) examined three scenarios for 2025: (a) a climate change scenario, in which two GCMs were used to predict changes in water supply; (b) a population growth scenario, in which water demand changed as a function of population growth, economic development, and efficiency improvements; and (c) a combined scenario, in which both climate and demand were changed. They found that climate change led to relatively small predicted increases in scarcity, while the two scenarios that included population growth had much more dramatic increases in the WTA scarcity indicator.

Alcamo et al. (2007) analyzed changes in water scarcity over the next 50 years for two different scenarios of population/energy use/greenhouse gas emission (the A2 and B2 scenarios from the Intergovernmental Panel on Climate Change), using two different climate models. They found that the percent of the global land area experiencing water scarcity by the Falkenmark and WTA indicators is likely to increase significantly and that this is mostly due to increases in population and water withdrawals, rather than climate-driven decreases in water availability (though this was an important factor for some locations). Their projections show that most of the world is expected to see a substantial increase in water withdrawals, driven mostly by growth in population, income, and industrial activity. Some areas are projected to experience a decrease in withdrawals, driven by increased efficiency and more favorable climatic conditions.

These two studies are no doubt not the last word on the matter. They both come primarily from the water use corner of the scientific community rather than from climate scientists. As a result, they are not as sophisticated on climate modeling as some of the other studies discussed above,

## Box 6.1. Climate Change and Water Availability in the Colorado Basin

The Colorado River serves as a vital resource for seven southwestern states and northern Mexico. Average precipitation over the basin is only about 400 mm, and the Colorado's flow is small compared to other North American rivers. The majority of the Colorado's flow originates as snowmelt in the relatively wet headwater mountains that make up a small fraction of the basin. The river then flows through a desert region where it loses water both to natural evaporation and to the many cities and irrigation districts that depend on it. Two large dams (Hoover and Glen Canyon) and many miles of canals are used to store and transport water.

Streamflow on the Colorado has been measured since the late nineteenth century. The measured flow data have been supplemented by extensive paleoclimatology work, in which tree rings have been used to reconstruct flows as far back as the ninth century. These results show that climate in the Colorado Basin has been highly variable at different time scales; this includes multidecadal periods with flows that are significantly lower or higher than the long-term average. For example, Meko et al. (2007) estimate that average flow during a 25-year period in the mid-twelfth century was about 15% lower than the long-term average. Such an extended period of lower-than-average flows calls into question the usefulness of the concept of "long-term average" and puts into perspective the recent drought, which has only lasted for about 6 years. A repeat of this type of extended dry period would have serious consequences in a basin in which every drop of the "average water availability" is already spoken for. (In fact, the Colorado Basin Compact that divided the entire river among the different states is based on an estimate of average flow that is now considered too high.)

If paleoclimatology has revealed that climate in the Colorado Basin has never been stationary, climate models have suggested that we may be on the verge of "an imminent transition to a more arid climate" in the region (Seager et al. 2007). Most global-scale and regional models appear to agree that average runoff in the Colorado is likely to decline, in large part because of higher ET due to warmer temperatures (NRC 2007). In addition, the relative role of snowmelt in generating streamflow is likely to decrease, resulting in changes in the seasonal timing of flow.

If indeed the climate of the southwestern US is transitioning to a drier state, should we think of this as a drought? As noted in section 5.4, droughts are best defined as deviations from the mean. But how long does a deviation have to last for us to start thinking of it as a new mean rather than a deviation? Perhaps a substantial climatic change of the type being forecast for the Colorado should be characterized as a "desiccation" or "increase in aridity" rather than a drought. On one level, this is purely a definitional question—yet underlying the semantics is the realization that water managers must adapt to a new reality rather than just pray for a speedy end to the drought.

and they may have a tendency to underestimate the potential impacts of climate change.

## Climate Change, Droughts, and Floods

Besides causing changes in overall precipitation for a given area, climate change is also expected to change the *distribution* of precipitation over time. One of the better-established predictions of climate science is that higher temperatures will lead to greater *variability* in precipitation (Trenberth et al. 2003, Groisman et al. 2005, IPCC 2007, Knapp et al. 2008). This will mean less frequent, more intense precipitation events, and could potentially exacerbate both flooding (during events) and water scarcity (in between events).

For example, the US Climate Change Science Program (USCCSP 2008) ranks as "very likely" the chances that this century will see "more frequent and intense heavy downpours and higher proportion of rainfall in heavy precipitation events" in North America. It should be noted that although this prediction of increased intra-annual variability (more intense events, longer periods between events) is essentially identical to a prediction of increased flooding, it is not identical to a prediction of increased drought, because drought requires an extended period of lower-than-average rainfall and is thus governed more by interannual variability. Increases in drought are ranked as "likely" only for certain parts of North America (USCCSP 2008). Still, the type of intra-annual variability that is considered very likely to be widespread (a longer period between rain events) can lead to increases in soil water stress and significant consequences for terrestrial ecosystems (Knapp et al. 2008).

Has the warming observed during the twentieth century already led to an increase in the frequency of intense precipitation events and resultant flooding? This question is being actively debated in the scientific community:

- Kunkel et al. (2003) found an increase in extreme precipitation in the US over the last several decades. However, they found that the frequency of extreme precipitation was also high around 1900, raising the question of whether the recent increase reflects a linear trend or simply a high point in a cyclical phenomenon.
- Groisman et al. (2005) examined precipitation records for much of the world and found an increase over time in the probability of intense precipitation in many regions.
- Milly et al. (2002) examined high flows in 29 large rivers that had long-term data (average record of 71 years). They found that 100-year floods occurred much more frequently in the latter half of the time period compared to the earlier half. They point out that the choice of large rivers and a high recurrence interval means that the floods are most likely affected primarily by precipitation, rather than by land use or hydrologic modification (e.g., dams). They also modeled the effects of fairly extreme climate change (a quadrupling of $CO_2$) on flood levels and found that large floods are predicted to occur much more frequently under this scenario.[2]

## Climate Change and Coastal Flooding

How is coastal flooding likely to be affected by climate change? Coastal flooding is caused not so much by extreme precipitation as by a storm surge of seawater driven by wind, typically during a hurricane. Another important factor in coastal flooding is sea level rise (SLR), which gradually decreases the elevation of coastal communities relative to sea level, making them more vulnerable to impacts from coastal storms.

Our question then becomes: To what extent have sea level and hurricane activity already changed, and to what extent are they likely to change in the future?

Sea level has been rising for the last 20,000 years or so, as Earth has warmed from the last Ice Age. Warming leads to higher sea levels for two primary reasons: the melting of land ice, which then flows into the oceans; and the warming of ocean water and the associated expansion in its volume. The average rate of SLR over the twentieth century was somewhere about 2 mm/yr. This is expected to increase during the

---

2. Specifically, the recurrence interval of what is currently a 100-year flood decreased in all but one of the basins; in 6 of the 29 basins, the predicted future recurrence interval was less than 10 years.

current century as a result of increased (human-caused) warming, but the year-to-year variability has made it hard to detect an acceleration at this point.

It is important to note that the actual rate of SLR experienced at a given location is a combination of the change in the water level of the ocean and the change in land level. This can lead to very high rates of *relative sea level rise* (RSLR) for areas where the land is sinking, such as the Mississippi Delta, where subsidence of unconsolidated sediment leads to a RSLR of about 10 mm/yr.

The hurricane story is more complicated. There is a great debate raging in the scientific community as to whether hurricane frequency and intensity have already increased (e.g., Emanuel 2005, Pielke et al. 2005, Landsea et al. 2006, Holland and Webster 2007). Complicating matters are the facts that there are natural cycles in hurricane intensity and that our methods for detecting hurricanes have improved dramatically over the last 100 years, which introduces a potentially large bias into our assessment of time trends.

There is not even agreement over whether warming is expected to lead to significant hurricane increases in the future. On the one hand, hurricanes get their strength from warm waters, so an increase in sea surface temperature should lead to higher wind speeds—and the damage potential of a hurricane increases as the cube of the wind speed. On the other hand, other atmospheric changes, such as increases in *wind shear* (the difference in wind speed and direction over small distances), may make it harder for damaging hurricanes to form.

## 6.2. Land Use Change

As discussed in Chapter 2, the quantity and quality of water in a river are closely tied to the nature of land cover and land use within the watershed. Changes in land use—such as those that humans have been implementing on a massive scale over the last century—can significantly affect the pathways of water flow and may alter the amount, timing, and quality of runoff. In this section, we discuss three types of land use change—deforestation, urbanization, and desertification—and their impacts on water resources.

### Deforestation

Vast tracts of forest are being cleared every day, especially in the wet tropics, for use as cropland and pasture. In addition, insects, diseases, and climate change are leading to loss of tree cover in certain regions. How does this affect the water budget and water quality of these areas? This question has proven to be both scientifically complex and politically controversial. On one hand, some have argued that increased timber harvesting should be used as a way to increase water availability (e.g., in the headwaters of the Colorado River). At the same time, others have pushed for reforestation of cut areas as a way to improve water quantity and quality (e.g., in the Panama Canal watershed). Which is correct? To answer this, we need to think about how forests affect the different parts of the hydrologic cycle.

*Transpiration* by trees is generally greater than transpiration by other plants, both because trees can tap more water sources with their deep, extensive root systems and because tree canopies have high roughness, which leads to more efficient transfer of moisture. Likewise, *interception* by trees is relatively high, especially when the canopy is dense, which leads to evaporation of rain before it can reach the soil. Thus, deforestation generally causes a *decrease* in ET and—assuming that there are no changes in precipitation—an *increase* in streamflow.

In general, *precipitation* will not be substantially affected by deforestation, although there are two important exceptions to this rule:

- *Cloud forests* or *fog forests* are ecosystems, generally found at relatively high altitudes, in which the forest canopy draws moisture from the air through the interaction of clouds and fog with the high surface area of the trees. Examples of such ecosystems include the redwood forests of coastal California and La Tigra National Park in Honduras, which serves as an important water source for the city of Tegucigalpa. Deforestation in these systems would lead to a reduction in the water that gets captured from the atmosphere.
- Deforestation over large areas can lead to a decrease in atmospheric moisture, due to the decrease in ET mentioned above. This in turn can lead to a decrease in precipitation, although this effect may be negligible depending on atmospheric circulation patterns. This effect is most relevant when we are looking at a large watershed in which much of the precipitation is generated from internal recycling of moisture between land and atmosphere within the watershed. A prime example is the Amazon basin, where models estimate that deforestation of 40% of the basin would result in a 12% decrease in precipitation; this would be somewhat counteracted by the decrease in ET mentioned

above, with the net result being a 4% decrease in streamflow (Coe et al. 2009).[3]

With the exception of these two cases, deforestation will generally lead to greater annual water yields from a watershed due to the decline in ET. This has been shown convincingly in both tropical and temperate watersheds (Bruijnizeel 2004, Likens and Bormann 2008). At the global scale, researchers have calculated that the conversion of forest to cropland and pasture has led to an increase in global runoff by 1800 km³/yr (Rost et al. 2008) or 3800 km³/yr (Falkenmark and Lannerstad 2005). (Of course, this has largely been counteracted by the decrease in runoff associated with consumptive water use in irrigation.)

However, there is another factor to consider: the ability of the watershed to store water and how this affects the *distribution* of runoff over time (Bruijnizeel 2004). In intact, forested watersheds, the porosity of the soil is quite high, and the soil can serve as a sort of "sponge," absorbing water during precipitation events and releasing it slowly. This is especially important in highly seasonal climates, where dry-season streamflow is completely dependent on the ability of watershed soils to store water and release it over many months.

When trees are removed and replaced with croplands or pasture, there is generally an associated degradation and compaction of the soil, including a loss of soil organic matter and a decrease in soil porosity and water-storage capacity. This means that more of the precipitation is likely to run off in the immediate aftermath of a rain event and less will be stored in the soil.[4] This will tend to cause both larger floods and more severe droughts. It should be noted, however, that the impact of deforestation on flooding is less significant for more extreme flood events, since these

---

3. The numbers given in the text are estimates for the basin as a whole. Coe et al. (2009) point out that this effect will vary substantially within the basin, with some sub-basins, such as the Rio Negro, experiencing substantial declines in streamflow (~10%) despite relatively modest deforestation rates (29%), while other sub-basins, such as the Tocantins, experience increases in streamflow (8%) despite massive deforestation (93%). This reflects the complexity of the processes that generate precipitation and runoff in Amazonia.

4. In regions where snowpack is important, there is an additional way in which deforestation leads to reduced water-holding capacity. Specifically, deforestation leads to reduced ability to retain snow, due to both the lack of shading and the absence of protection against avalanches. This leads to earlier, higher snowmelt flows and reduced summertime flows.

are caused by extreme levels of precipitation that far exceed the water-holding capacity of even the most highly forested watersheds.

The net effect of deforestation on dry-season flows is dependent on the balance between the two competing effects discussed above: the overall increase in annual flow (due to the decrease in ET) and the decreased soil storage. Where the former dominates, dry-season flows will increase, while in areas where the latter effect dominates, dry-season flows will decrease (Bruijnizeel 2004). These effects have thus proven to be quite site-specific and difficult to predict (Calder 1999).

The real problem with deforestation, though, is probably water quality rather than quantity. Several factors are at play. Cleared areas that are left barren of vegetation can release large amounts of nutrients that are no longer being taken up by growing plants. Exposed soils are susceptible to erosion and carry with them sediment-bound nutrients and contaminants. The rapid movement of water through a deforested watershed short-circuits the natural filtration provided by forest soil and vegetation. In addition, the grazing or agricultural activity that replaces the forest usually introduces new sources of pollutants, including pathogens from animal waste and agricultural chemicals like fertilizers and pesticides (see Chapter 10). These pollutants tend to ultimately be released to nearby streams and rivers.

Can reforestation of cleared areas solve the water problems associated with deforestation? Probably not right away. It can take some time for trees to grow enough to increase ET to pre-clearing levels. More important, it can take even longer for soils to recover their moisture-holding and pollutant filtration capacity, especially if the area was heavily grazed or the soils were heavily eroded.

## Urbanization

The vast majority of deforestation involves conversion to agricultural, rather than urban, land. Still, in both developed and developing countries, there also is increasing conversion of land to human habitation, especially in urban conglomerations and their associated suburbs. What is the effect of this urbanization phenomenon on the quantity and quality of runoff? We explore this topic in greater detail in section 8.6, where we discuss various aspects of the ecosystem degradation that is typical of urban streams. For now, I provide a brief summary of the impacts of urbanization on water resources.

Similar to deforestation, conversion of undeveloped or agricultural land to urban land uses causes a reduction in the capacity of the watershed to store water. This leads to increased "flashiness" (a more rapid response to storms), which implies greater risk of flooding (and potentially also lower dry-weather baseflows).

As was the case for deforestation, it is likely that the effects of urbanization are diminished for floods with longer recurrence intervals and larger spatial scales. The former has to do with the fact that extreme floods are by definition unusual events that overwhelm the capacity of soils to absorb water; the additional increment contributed by urban areas becomes less significant. The latter (the limited relevance of urbanization for large-scale flooding) has to do with two factors: in large watersheds, you are less likely to see a high degree of urbanization; and flooding in large rivers is more affected by the relative timings of flow from different tributaries than by the timing of flow within a single tributary.

In addition to the problems of flooding and decreased baseflow, urbanization can also create water scarcity through the degradation of water quality. Many rivers, especially in developing countries, are unfit for human use because of the inputs of municipal and industrial waste.

## Land Use Change and Coastal Flooding

How has coastal flooding been affected by land use change? The processes of deforestation, agricultural expansion, and urbanization discussed above have played out in coastal as well as inland areas, although in somewhat different ways. Population growth has been dramatic in the coastal zone, with almost 10% of the world's population now living within 10 km of the ocean (Small and Nicholls 2003). This growth has put more people into flood-prone areas and has resulted in the widespread degradation of coastal ecosystems—a process that, in many cases, has increased the magnitude of flooding. Two dramatic examples illustrate the patterns: the Indian Ocean tsunami in 2004 and Hurricane Katrina in 2005. In the former, there is some evidence that areas where mangrove forests and coral reefs had been destroyed tended to experience more flood damage than areas where these protective ecosystems were intact, although this assertion has been disputed (Venkatachalam et al. 2009). In the latter case, the devastation caused by Hurricane Katrina was exacerbated by the loss of vast areas of coastal marsh that once stood between New Orleans and the Gulf of Mexico (Costanza et al. 2006). (The reasons for marsh loss in coastal Louisiana are much disputed but are in some way related to anthropogenic changes in the landscape [Turner 1997, Day et al. 2000].)

## Desertification

The land use changes discussed above involve relatively simple and well-understood causal linkages, in which human changes to the land lead to changes in water quantity and quality. Desertification, in contrast, is a more complex and controversial subject in which separating cause and effect—or even agreeing upon a definition of the issue—has proven remarkably difficult. Fundamentally, desertification involves the expansion of desert (land with low precipitation rates and sparse vegetation cover), but there is much dispute about the relative importance of climatic and nonclimatic causes.

One perspective sees desertification as a significant and growing problem caused in large part by direct human activity. In this view, decreases in rainfall may initiate the process, but desertification is driven primarily by anthropogenic impacts such as overgrazing, soil compaction, and removal of vegetation. These human changes lead to a degraded landscape that has a poor ability to hold water, so when the rains do come, they wash off in flash floods and don't benefit the desiccated soils; as a result, the ability of the land to support vegetation is diminished, leading to a vicious cycle of landscape degradation. In addition, vegetation loss can increase the albedo (reflectivity) of the land surface, which can cause a decrease in rainfall, further reinforcing the cycle (Charney et al. 1975).

A contrasting viewpoint sees climatic variability as the main driver of changes in precipitation and vegetation in arid regions, and perceives no strong global trend of increasing desert area. This paradigm emphasizes the high interannual variability in precipitation that is typical of arid and semiarid climates, and sees multiyear droughts as part of the natural cycle—though anthropogenic climate change may also play an important role in changing precipitation patterns. By this logic, land cover is naturally dynamic, with vegetation decreasing during dry spells and recovering during wetter periods.

Of course, there is ample room to take pieces from both perspectives. Climatic variability may set the stage for changes in water, land, and vegetation, but human land and water management can also have significant impact—either positive or negative—at least at the local scale and probably beyond. In fact, a better understanding of climate variability and ecosystem response can lead to better management. For example, the "carrying capacity" of a rangeland is best thought of as a dynamic quantity that will increase in wetter periods and decrease during dry years.

## Box 6.2. Drought and Desertification in the Sahel

The Sahel region of Africa, on the border between the Sahara Desert to the north (annual rainfall <200 mm) and the subhumid savanna to the south (annual rainfall >600 mm), is a region where rainfall varies dramatically in space and time, leading to equally dramatic variation in vegetation and carrying capacity for livestock. The spatial variation in average precipitation can be seen by examining Figure 3.2 (Chapter 3), which illustrates the strong gradient of decreasing rainfall from south to north. The seasonal variability in rainfall can be seen in NASA's animation of global monthly rainfall (precip.gsfc.nasa.gov/gifs/gpcp_rain_79-99.qt), which shows that the Sahel gets rain only during June–October, when the intertropical convergence zone (ITCZ) moves north into the area. But perhaps even more important is the interannual variability in rainfall, illustrated in Figure 6.1.

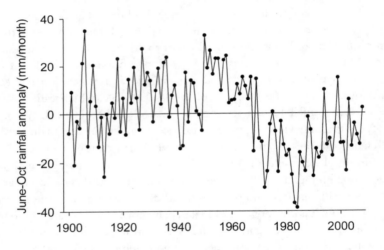

*Figure 6.1.* Sahel June–October rainfall anomaly (difference from 1900–2008 mean). Data source: jisao.washington.edu/data_sets/sahel/.

This figure shows clearly the multidecadal drought (or series of droughts) that have afflicted the Sahel since the late 1960s (with some recent alleviation) and has led to severe impacts on vegetation, livestock, and people, including serious famine. The desertification debate alluded to above has been playing out in a major way in the Sahel, with scientists unable to provide definitive answers to some of the most basic questions about this drought:

- Has there really been a decrease in rainfall over the Sahel? The lack of a consistent, robust meteorological network in this region has led some scientists to raise the question of whether the patterns seen in Figure 6.1 are representative, though the answer appears to be yes.
- Is the decrease in precipitation driven primarily by climatic factors or by land use change? The main climatic factor that has been invoked is sea surface temperature: when temperatures increase in the Southern Hemisphere relative to the Northern Hemisphere, the ITCZ moves farther south, bringing reduced rainfall to the Sahel. In contrast, the land use hypothesis argues that loss of vegetation has led to decreased atmospheric moisture (through decreased ET) and increased albedo, both of which contribute to decreased rainfall.
- Is the drought part of a long-term cycle of oscillations in rainfall, or is it a transition to a new, drier state (a "desiccation"), caused by either climate change or land use change?
- Is vegetation loss in the Sahel caused mostly by reduced rainfall or by anthropogenic causes, specifically overgrazing and the expansion of human settlements?
- Will vegetation recover naturally if and when rainfall returns, or will it require a concerted human effort aimed at changing harmful land use patterns and instituting better soil and water management practices?

For different perspectives on this issue, see Hulme 2001, Petit et al. 2001, Dai et al. 2004, see Held et al. 2005, Herrmann et al. 2005, Hein and de Ridder 2006, Chappell and Agnew 2008, and Govaerts and Lattanzio 2008.

## 6.3. Conclusion

We live in an era of rapid environmental change. As discussed in this chapter, changes in climate, land use, and population are all important factors in our water problems. These factors differ in some important ways. Changes in land use and population already exert large influences on water issues, while the effect of climate change is mostly, though not exclusively, a concern about the future. Scientific uncertainty tends to be most significant for climate change, though there is some uncertainty surrounding all three issues. Land use change is a local phenomenon that is playing out around the globe, while climate change is a global issue whose effects will be felt in local alterations in water availability. As illustrated by the Sahel example, land use and climate issues interact in significant ways. Both phenomena are ultimately driven by population growth, which also has its own direct effects on water demand.

What these three factors have in common is that they pose significant challenges to future water availability in a world already facing scarcity. Yet there is much we can do to improve water management and create a better future in the face of these challenges. We turn now to examining the technologies that are available to deal with scarcity.

# 7

## Soft and Hard: Technologies for Sustainable Water Management

In addressing the challenges of water scarcity in a changing world, we need to think deeply about what tools are most appropriate. Each era of water management is strongly associated with certain technologies—those that at that point in time were both technically feasible and compatible with the prevailing mindset. For example, as discussed in Chapter 1, water management in the twentieth century came to be defined by dams, canals, and wells.

Where do we stand now, at the beginning of the twenty-first century? What new technologies are emerging? What old technologies are still appropriate? Each section of this chapter will discuss in some detail a technology, or group of technologies, for addressing water scarcity. The final section will summarize and compare the different options.

### 7.1. Large Dams

As noted in section 1.1, large dams are the preeminent symbol of the hard path, and the focus of very strong feelings both pro and con. Some of the most important organizations involved in the dam debate include the following:

- The International Commission on Large Dams (ICOLD) is an association of dam builders and operators that, among other activities, establishes guidelines for dam construction and operation, and maintains a registry of large dams based on submissions from member countries.

- The International Hydropower Association is a trade group representing the hydropower industry.
- International Rivers (formerly the International Rivers Network) is an environmental NGO that works against large dams worldwide.
- The Narmada Bachao Andolan (Save the Narmada) is a local group that has worked to try to stop the Sardar Sarovar Dam on the Narmada River in India and has received worldwide attention for its dramatic methods and passionate leadership.
- The World Commission on Dams (WCD) was established in 1998 by the World Bank and the International Union for Conservation of Nature (IUCN) in an effort to reach some compromise in the battle between dam advocates and dam opponents. The 12 commissioners who made up the WCD were influential people from both sides of the debate. Despite some predictions that they would not be able to reach consensus, the WCD published a final report (WCD 2000) with strong conclusions and guidelines for future dam projects.

Despite the polarization of the debate, we will try to evaluate, as objectively as we can, the costs and benefits of large dams. In doing so, we draw heavily on the data compilation and analysis carried out by the WCD, specifically their assessment of the performance of large dams from technical, financial, environmental, and social perspectives.

## Technical and Financial Performance

One focus of the WCD was on evaluating completed dam projects against the performance projections that were made before construction began, using several criteria:

- Cost: The WCD found that average cost overruns for large dams were 54%. This compares unfavorably to World Bank analyses of fossil fuel plants (6% overruns) and development projects in general (11%).
- Timing: Most large dams were completed within a year of schedule, although some lagged behind.
- Benefits: There was great diversity in dam performance, with some projects providing fewer benefits than expected and others providing more. On average, all categories evaluated (irrigation, water supply, hydropower) underperformed relative to

expectations, though hydropower dams did the best on average, with the overperformers compensating for the underperformers.

These results lead one to question whether *cost-benefit analyses* (CBAs) for dams can be trusted. CBAs, which are discussed in detail in Chapter 12, are the primary tool used for evaluating whether the benefits of a dam (e.g., the hydropower that it will provide) outweigh the costs (financial, environmental, social). The integrity of the CBA process depends on an ability to predict both costs and benefits in a reasonably precise and unbiased manner. The WCD results summarized above suggest that this is often not the case, even for the direct costs and benefits that should be relatively straightforward to evaluate.[1] This may be in part because constructing a large dam is inherently a very complex task in which every site poses new challenges and in part because of the large political and financial pressures to inflate benefits and underestimate costs in order to make marginal projects appear favorable.

## Dam Safety and Maintenance

Dams hold back huge volumes of water, and when they fail—as all human-made structures may—they can cause extensive damage. McCully (2001), in his anti-dam book *Silenced Rivers*, documents 48 cases of dam failures that each killed at least 10 people. By far the greatest devastation was caused by a multiple-dam failure in China in 1975, when unusually high rainfall led to the overtopping and destruction of the Banqiao and Shimantan dams, along with up to 62 smaller dams. At the time, this event was hidden from the outside world, but a later Human Rights Watch report (HRW 1995) estimates that 85,000 people were killed by the initial flood wave and another 145,000 people died from starvation and disease in the weeks following, as floodwaters covered thousands of square kilometers.

Dam safety is highly dependent on two factors that are unique to each site and hard to get good data on: the geology that supports the dam structure and the hydrology that determines how much water it must be able to withstand. The WCD points out that our record on dam safety seems to be improving over time, with a failure rate of <0.5% for dams built after 1950, compared to 2.2% for dams built before then.

---

1. Environmental and social costs, such as disruption to ecosystems, are much harder to quantify and are often not included in CBAs; see below and Chapter 12.

An additional safety factor to consider is the issue of *reservoir-induced seismicity*. There is substantial evidence (McCully documents 32 cases) that the pressure of massive amounts of water, combined with water's lubricating effects, can lead to intensified earthquake activity when dams are built in seismically active areas. Shortly after the 2008 earthquake in Sichuan Province, China, which killed about 90,000 people, American and Chinese scientists suggested that the quake may have been caused, at least in part, by the 320 million tons of water in the nearby Zipingpu Reservoir (LaFraniere 2009). A modeling study by Ge et al. (2009) confirmed these suspicions, calculating that the reservoir accelerated the quake by tens to hundreds of years.

Dams have a limited life span before they need significant repairs or decommissioning. The 1960s peak in dam-building in the US means that, in the 2010s, many dams (both large and small) will be reaching the 50-year mark, a point at which they may require significant investments to maintain their safety and function. The American Society of Civil Engineers gives dams in the US an overall grade of D (ASCE 2009). Are the benefits provided by dams just a temporary feature, with large costs (construction and decommissioning) on either end, or can we find ways to keep this generation of dams functional for many years to come?

## Ecological Impacts

Dams can cause serious environmental impacts, including river fragmentation, prevention of fish migration, sediment trapping, and water quality problems. These are discussed below. In addition, dams lead to changes in river flow regimes, an issue that is discussed in detail in section 8.2, and to greenhouse gas emissions, discussed in section 11.1

Perhaps the most overarching impact of dams on river systems is *fragmentation*: the alteration from a free-flowing, integrated ecosystem to a series of disconnected reaches separated by barriers. Nilsson et al. (2005) conducted a global analysis of fragmentation by large dams for the world's 292 largest rivers. They found that 190 (65%) of these river systems had some degree of fragmentation, with 173 (59%) having dams in tributaries and 153 (52%) having dams in the main channel. The unfragmented rivers were mostly in areas with little economic activity, particularly in tundra zones and in Australasia.

One consequence of fragmentation is the inability of fish, especially migratory ones, to move through the river system. Dams of almost any size create a barrier to fish movement in the upstream direction. In addition, large reservoirs and dams can create a barrier to downstream fish movement, since travel through a reservoir requires more energy and time than being carried downstream by the current, and travel over a dam spill-way (or through hydropower turbines) can be deadly. Various structures can be installed to allow fish to pass around a dam, ranging from the traditional *fish ladder*, in which fish move up from one small pool to another, to *nature-like fishways*, which are designed to mimic natural stream channels. Fish passage structures are generally designed to pass a specific species or group of species, and their effectiveness is not always high. For smaller streams, structures other than large dams can also cause fragmentation. In particular, even small weirs a couple of feet high can pose passage issues for some fish. Culverts used for road crossings can also lead to fragmentation, especially when flows are low, for two reasons: fish may be unable or unwilling to traverse the culvert due to shallow water depths or darkness, and/or there may be a significant drop in height from the outlet of the culvert to the downstream section of the stream due to erosion.

Fragmentation by dams also leads to *sediment trapping*. Rivers naturally carry sediment downstream, sometimes in massive volumes. When a river is interrupted by a section of quiescent water behind a dam, it will deposit its sediment. This may cause both loss of storage volume in the reservoir and disruption of downstream sediment dynamics, often leading to erosion of downstream river channels. The WCD reports that the rate of reservoir sedimentation averages 0.5–1.0% of reservoir volume per year, but is highly dependent on the location of the dam.

Dams can also lead to several water quality problems. First, the bottom waters of reservoirs (and natural lakes) are often low in oxygen, high in sulfide, and relatively uniform in temperature year-round. Dams that have intakes at depth will release these waters downstream, with significant harmful effects on organisms. In addition, reservoir bottom sediments are hot spots of mercury methylation, the bacterial process that converts mercury to methylmercury, a form that is more toxic to people. Third, when urban or industrial areas are submerged in the rising waters of a new reservoir, the myriad pollutant sources that are found in these areas may leach contaminants into the waters of the reservoir. Lastly, the long residence time of water in reservoirs can lead to eutrophication problems. This in turn leads to a change in the nutrient ratios that reach the sea, since reservoirs preferentially trap silicon, primarily through incorporation into algae and burial in sediments. This contributes to low Si:N (silicon:nitrogen) ratios in estuaries and other downstream ecosystems, which causes changes in algal communities (a decline in diatoms) and tends to increase the incidence of harmful algal blooms.

## Social Impacts

Arguably the greatest negative impacts of large dams come when homes and property are flooded by the rising waters of a reservoir. The WCD estimates that 40–80 million people worldwide have been directly displaced from their homes by large dams. Although resettlement programs are supposed to provide new homes and livelihoods to dam refugees, these programs have mostly been woefully inadequate. Thayer Scudder, perhaps the foremost expert on dam resettlement projects (and a member of the WCD), has documented the devastating effects of displacement on individuals and communities. He notes that of 50 large dams surveyed, only 2 achieved the goal of improved living standards for the majority of resettled people, while 31 resulted in significantly worsened conditions for the majority (Scudder 2005). Those displaced by the dam are rarely allowed to share fully in the benefits that the dam brings, such as jobs, electricity, water supply, fishing, and tourism.

In addition to those directly displaced by dams, other groups of people may lose their livelihoods as a result of the ecological changes caused by the dam, both upstream (the flooding of land for the reservoir) and downstream (the alteration to the river flow, especially affecting fishers and floodplain farmers).

Other social impacts of dams can include

- submergence of sites of archaeological, cultural, or religious significance;
- influxes of construction workers from other regions, leading to overcrowding and social conflict; and
- increases in certain diseases, such as malaria (due to increased mosquito breeding grounds), schistosomiasis (due to increased snail habitat), mercury poisoning (due to the mercury methylation discussed above), and alcoholism (due to new settlement patterns and breakdown of social structures). Malaria and schistosomiasis are discussed in more detail in Chapter 9.

## Positions on the Dam Debate

Large dams clearly bring benefits as well as costs. Dam proponents argue that the benefits generally outweigh the costs and that the social and environmental impacts can be identified and mitigated through appropriate measures such as resettlement, wise choice of sites, and so on. They see dam projects as unique engines of human development. Dam opponents, on the other hand, argue that the benefits of dam projects tend to be overstated, while the social and environmental costs are often poorly quantified or ignored completely. They are skeptical of mitigation measures and emphasize the inequity in the distribution of costs and benefits.

I believe that where one stands on this debate depends in part on one's reading of the available evidence, but in part also on one's underlying, intuitive feelings about three issues:

*Is nature good or bad?* Dam proponents tend to see nature as needing taming and improvement to serve humanity, while dam opponents tend to have an idyllic view of humans living in harmony with nature.

*Can we quantify social and environmental impacts?* Dam proponents tend to see economic cost-benefit analysis as an appropriate tool for weighing the different factors, while dam opponents tend to argue that certain impacts can't be quantified.

*Medium-term or long-term thinking?* Dam proponents tend to emphasize the vital needs of the present, while dam opponents tend to think of dams as structures of limited lifetime that will not meet long-term needs.

Regardless of one's position on this debate, it is important to keep in mind that not all dams are the same. The social and environmental impacts of the worst dams are much higher than those of the best dams. Likewise, the benefits provided by the best dams are much higher than those provided by the worst. One challenge facing us as we move forward is to better identify those factors—hydrologic, geographic, geologic, social, legal, financial—that determine whether a particular dam project ends up being one of the best or one of the worst. We then need to use that information to make sure that the worst projects don't get constructed, or that they get modified to reduce their worst impacts. The World Bank and other financing institutions have started to put in place environmental and social safeguards to ensure that their projects meet basic criteria and don't end up on the list of worst dams. However, multilateral lending institutions like the World Bank are now playing a smaller role in financing dam projects (in part because of the bad publicity they have received). Their place is being taken by bilateral finance agencies, such as export credit agencies and national development banks, which tend to have weaker environmental and social safeguards (Caspary 2007). China, which has very little in the way of safeguards, is also playing a larger and larger role in dam-building in developing countries.

## 7.2. Small Dams

Dams that are less than 15 meters high are not included in ICOLD's dam registry, so we lack reliable numbers on how many there are worldwide.

In the US, the National Inventory of Dams, maintained by the Army Corps of Engineers, lists about 79,000 dams greater than 2 meters high, of which only ~6400 qualify as large. There are undoubtedly many more thousands smaller than 2 meters in height.

Smaller dams in the US are often used for somewhat different purposes than larger ones. They are more likely to be used for recreation and often serve as a local water supply for irrigation or firefighting. They are also much more likely to be privately owned.

How do small dams compare to large ones in terms of the benefits provided as well as the social and environmental impacts? On the one hand, small dams—even a large number of them put together—clearly can't provide the kinds of multiyear water storage and impressive hydropower that large dams can. Likewise, small reservoirs are susceptible to drying out in droughts, and small dams can be washed out in floods. In addition, small dams tend to be poorly regulated and can pose significant maintenance and safety problems, especially when their owners have abandoned them or gone out of business.

On the other hand, there are certain advantages to operating at a smaller scale. Construction and decommissioning are cheaper and less likely to attract corruption. Benefits are more likely to stay in the local area. Silt can be removed from ponds and used productively. Environmental impacts—even cumulatively—tend to be lower. Perhaps most important, small infrastructure tends to encourage a locally based water economy that emphasizes living within limits and adapting to environmental constraints.

## Dam Removal

There has been great interest recently in the "selective removal of dams that don't make sense," as a publication from the environmental group American Rivers puts it (American Rivers 1999). This movement is driven by both environmental and economic motivations. Environmentally, the goal is to restore river ecosystems by reversing the various ecological impacts associated with dams (mentioned above and discussed in more detail in Chapter 8). Economically, many of the dams that are being removed are no longer serving their original purpose, are in need of significant repair, and may not even have identifiable owners; dam removal is often the cheapest option. It is estimated that 85% of US dams will be near the end of their operational lives by 2020 (Doyle et al. 2003). Likewise, many hydroelectric dams built in the 1950s and 1960s are reaching the end of their operating licenses from the Federal Energy Regulatory Commission (FERC); since FERC is increasingly requiring stringent and expensive environmental mitigation as part of the relicensing process, dam removal may be the most financially attractive option.

Dam removal has taken place around the world, but has perhaps been best documented in the US, where American Rivers reports that at least 500 dams have been removed, most of which have been small (median height ~5 m, largest = 49 m). Plans are under way for the removal, beginning in 2012, of two large dams (33 m and 64 m) on the Elwha River in Washington State. Dam removal advocates are pushing for still larger targets, including four dams on the lower Snake River in Washington and, perhaps quixotically, the Glen Canyon Dam in Arizona.

How well does dam removal accomplish its goal of ecosystem restoration? The evidence suggests that flow regimes and fish passage are restored to their pre-dam conditions relatively reliably and quickly (Bednarek 2001). However, the significance of these changes for ecosystem restoration depends on other impacts that may still exist, including hydrologic (e.g., water withdrawals), physical (sediment loads, channelization), water quality (pollutant loading), and biotic (overfishing), as well as the presence of other dams upstream or downstream on the same river. In addition, questions remain as to how river geomorphology (channel shape) readjusts to the new conditions presented by dam removal (Wildman and MacBroom 2005).

Here are some of the issues that need to be addressed in planning for dam removal:

- Sediment dynamics: Large volumes of sediment tend to accumulate behind dams, and thought needs to be given to the fate of this sediment following dam removal.
- Contaminants: Sediments need to be tested for contaminants to ensure that dam removal doesn't redistribute pollutants. The infamous 1973 removal of the Fort Edward Dam on the upper Hudson River resulted in the unwitting mobilization of sediments containing tens of thousands of kilograms of PCBs, significantly complicating the eventual environmental cleanup.
- Opposition: There may be local opposition to dam removal from citizens who use the reservoir or see the dam structure itself as historically significant.
- Watershed perspective: Dam removal projects need to be prioritized based on how many river miles they open up for fish

passage, as well as whether other environmental impacts are likely to mitigate the success of the project.

- Permits: Obtaining federal and state permits for dam removal can be a time-consuming and frustrating process.

## 7.3. Canals and Water Transport

As discussed in Chapter 3, the world's water resources are not evenly distributed. Some densely populated areas (or arable land) have little water nearby, while other sparsely populated areas are water-rich. It is natural, then, to think about solving water scarcity problems by moving the water to where it is needed. Indeed, as noted in Chapter 1, long-distance transport of water using canals or other methods is a key feature of twentieth-century water management. These water transfers are most common within countries (Box 7.1), though they sometimes take place internationally as well.

There are several problems, both economic and environmental, with this approach:

- Water is heavy relative to its value. This means that transporting water large distances is a fairly expensive proposition and is affordable for only the most highly valued (or most highly subsidized) uses of water.
- Water transfers do not appear to provide a solution to regional water scarcity, as experienced, for example, in the Middle East and North Africa (MENA), where most countries are already experiencing scarcity by both the Falkenmark and the WTA indicators. Water imports for these countries would have to come from relatively far away, which would be economically prohibitive as well as politically problematic and environmentally harmful. Although ideas of a pipeline from Turkey or large oceangoing water-filled bags have been floated (so to speak), they are unlikely to be implemented.
- Long-distance water transport provides ample opportunity for loss of water to both evaporation and seepage.
- Where water is transferred between basins, there are significant changes to flow in both the giving and the receiving basins.
- Canals often pass through pristine areas, where canal construction can have significant environmental impacts.
- Long-distance transport of water can lead to movement of non-native aquatic species and loss of native species.

- Water transport allows and encourages the development of cities and farms in places that do not have the water to support them. This sets in place a growth dynamic that necessitates ever-increasing water transport to satisfy ever-growing demand. It also results in a system where the water supply for large cities is vulnerable to disruptions, whether physical (earthquake damage to canals), economic (increased cost of transporting water), or social (terrorist attacks on water transport infrastructure).

## 7.4. Virtual Water Trade

For many water-scarce countries, it may make more economic sense to import water-intensive foods and products rather than water itself. The water embedded in these products has been referred to as "virtual water," and the trade in these products thus represents virtual water trade.

Many MENA countries, for example, have apparently recognized that they don't have enough water to be self-sufficient in food and are currently importing large quantities of grains. Transporting a ton of grain is much cheaper than transporting the hundreds of tons of water that it takes to grow the grain. And given the large quantities of water required for growing food, importing even a modest amount of food can free up a substantial amount of water for other uses.

Note that this strategy for alleviating water scarcity is generally not articulated explicitly by the countries involved. In several MENA countries, such as Egypt, it is politically untenable to publicly acknowledge the infeasibility of water self-sufficiency.

Besides alleviating country-level or regional water scarcity, virtual water trade can potentially serve three other functions. First, it can increase global water use efficiency by growing crops (and making industrial products) in countries where they can be made with less water. Thus, for example, shifting cereal production to favorable climatic conditions (less evaporative demand) can lead to an overall decrease in the amount of water needed to feed the world. Second, virtual water trade can shift agricultural water use from blue water to green water. That is, countries whose dry climates require irrigated food production can import food grown in more humid countries using rain-fed agriculture. This may not necessarily change the total amount of water required but could free up blue water for other uses that require it (e.g., drinking water). Third, by lessening a country's dependence on local sources of water, virtual water trade may be able to alleviate tensions over transboundary water resources.

## Box 7.1. South-to-North Water Transfers in China

Like the US, China is large enough that it contains regions with very different climatic conditions (see Figures 3.2–3.5). While southern China receives average annual precipitation of 1000–1500 mm, much of northern China receives only 600 mm, dipping to as low as 35 mm in the desert northwest. Despite this, the North has almost half the population of the country (including Beijing and many other large cities) and much of the agriculture. As a result, North China (defined here as the watersheds of the Huang, Hai, and Huai and the northern endorheic region) has a per capita water availability of 704 $m^3$ person$^{-1}$ year$^{-1}$, compared to 3230 in the rest of the country (Shao et al. 2003). The Huang He (Yellow River) in the North is fully utilized and no longer reaches the sea much of the year, while the Yangtze (Chiangjiang) River in the South still has abundant flows (though it has certainly suffered its share of ecological damage, including the massive Three Gorges Dam). Agriculture in the North has been surviving on unsustainable groundwater use, and Beijing and other northern cities are in real danger of water scarcity. Unregulated pollution of surface and groundwater is compounding the problem.

As a result, China has begun implementing an ambitious and controversial plan to move water across the country from south to north, an idea that has been under consideration for more than 50 years, since Mao first commented, "If possible, lending some water [from south to north] would be good." There are already seven interbasin water transfers operating in China, but they traverse relatively short distances and move relatively small amounts of water. The new diversion project will consist of three separate routes that, if fully implemented, could potentially transfer as much as 45–60 km³/yr, augmenting northern China's water resources by some 15–20%.

The Eastern Route, which is currently under construction, will divert water from the Yangtze to serve the northern city of Tianjin and nearby farmers. The water will travel 1150 km, largely using the centuries-old Grand Canal, and will need to be lifted 65 m by pumping stations before it passes under the Huang (a difficult technical challenge in itself) and travels by gravity the remainder of the way. Water quality is a severe challenge for this route, given the high loads of untreated wastewater that are discharged into the Grand Canal; a possible side benefit of this project is that China may finally get serious about treating sewage and industrial waste.

The Central Route, which is also under construction, will divert water from a reservoir on the Han River (a tributary of the Yangtze), primarily to serve Beijing. Water will travel by gravity along the 1240-km-long route, but construction challenges are formidable, including both digging the canal itself and making improvements to the originating reservoir and other reservoirs that will store water en route. Water from the northern portion of the Central Route (running 307 km from Hebei province) was delivered to Beijing in time for the 2008 Olympics.

The Western Route is still in the planning stages, but it is anticipated that it would transfer water from the headwaters of the Yangtze into the headwaters of the Huang in order to serve agriculture in northwestern China.

The South-to-North Water Transfer is the epitome of the hard path. It will be an impressive engineering feat, but will create numerous environmental and social problems. Some of the most important issues are the downstream effects of reduced flow in the Yangtze, especially an increase in eutrophication; the displacement of hundreds of thousands of people along the route of the canals; unanticipated technical difficulties that may increase the costs of construction and operation; increased incidence of schistosomiasis (see Chapter 9); and soil salinization problems.

Could North China's water-scarcity problems be solved with the soft path? The answer is unknown, but several approaches could be tried that in combination would likely go a long way in that direction: price increases and efficiency measures for both urban and agricultural uses; treatment of wastewater to prevent contamination of water sources; and rethinking the growth of population and water-intensive activities in an arid region.

To what extent do empirical data on virtual water trade support the theory outlined above? Several groups have attempted to calculate virtual water flows between countries by combining two types of data: global trade data, especially for agricultural products; and data on the amount of water required to produce these products in both the importing and exporting countries. Their results have addressed the two most important questions related to virtual water trade.

*Which countries are most heavily engaged in virtual water trade?* As expected, the big virtual water exporters are the grain–exporting countries, such as the US, Australia, and Argentina, while virtual water importers include Saudi Arabia, Jordan, Japan, Italy, Germany, and Mexico. As originally noted by Allan (1998, 2001), the water–scarce MENA countries are almost all virtual water importers. Yang et al. (2003) note that countries with a per capita water availability below 1500–2000 $m^3$ person$^{-1}$ yr$^{-1}$ tend to be virtual water importers, with the amount of import increasing dramatically as countries move farther below this threshold. However, Yang et al. (2006) calculate that 68% of global virtual water import is in non-water-scarce countries and conclude that "water scarcity has a relatively limited role in shaping the global virtual water trade flows." Kumar and Singh (2005) and Roth and Warner (2008) point out that a country's status as a grain importer or exporter is affected by many other factors besides water, such as arable land availability and national food policy. It is probably safe to conclude that water is the driving force behind food trade only in the most water-scarce countries—and then only to the extent that they can afford to import food.

*Does virtual water trade result in global water savings?* The consensus answer to this question seems to be "yes," with calculated net global savings of 352 km$^3$ yr$^{-1}$ (Hoekstra and Chapagain 2008), 337 km$^3$ yr$^{-1}$ (Yang et al. 2006), or 164 km$^3$ yr$^{-1}$ (de Fraiture et al. 2004; cereals only). On the whole, we are growing agricultural products in locations that are relatively water-efficient and exporting them to locations where it would take more water to grow the same products. Data also show that much of the virtual water trade consists of food that was grown using green water and was exported to countries where blue water would have been used to grow it (Yang et al. 2006).

Transfer of virtual water certainly takes place within countries as well as between countries, but there has been relatively little analysis of these transfers. An exception is a recent analysis for China (Box 7.2).

There has been some theoretical opposition to the virtual water concept, with some researchers criticizing it as "redundant" (e.g., Merrett 2003). In addition, many people outside of the water policy circle are taken aback by the suggestion that solutions to water scarcity lie in transporting food around the world. This opposition arises, I think, from four legitimate concerns:

- Global food trade flies in the face of the "local food" movement, which promotes the environmental, nutritional, and social value of locally grown food.
- In importing countries, there can be nervousness about being dependent on other countries for such a basic necessity. In addition, there is often an aesthetic and cultural attraction to preserving a farming tradition, even when the water to support that tradition is no longer available.
- In exporting countries, there can be opposition to the idea of "selling our water." Indeed, it seems that for some countries,

---

### Box 7.2. North-to-South Virtual Water Transfers in China

Regional food trade within China provides an example of virtual water transfers at the intra-country scale—an example that shows how virtual water transfer can, in some instances, exacerbate, rather than solve, water scarcity.

South China, with its abundant water resources, was traditionally the country's breadbasket. However, over the last 20 years or so, as the South has become increasingly industrialized, agricultural production has shifted to the North, where fertile land and abundant sunshine provide ideal conditions for farming. Except for the absence of water. As noted in Box 7.1, agriculture in the North has been living on unsustainable groundwater use and is eagerly anticipating the South-to-North Water Transfer.

Hoekstra and Chapagain (2008) calculate that North China[a] annually exports 51.6 km$^3$ of virtual water to the South, primarily in the form of grains and vegetables. They note the absurdity of transporting so much virtual water from an arid region to a wet one and point out that eliminating this virtual water trade would provide as much water to the North as will be provided by the South-to-North Water Transfer. Of course, eliminating this trade would cause severe socioeconomic disruption. But are there ways to move toward an economy in which water availability plays a more prominent role in determining the patterns of agricultural and industrial production?

[a] North China is here defined slightly differently than in Box 7.1, as comprising the provinces of Beijing, Tianjin, Shanxi, Inner Mongolia, Liaoning, Jilin, Heilongjiang, Hebei, Henan, Shandong, Anhui, Shanxi, Gansu, Qinghai, Ningxia, and Xinjiang.

---

like India, the export of food is based on unsustainable and environmentally harmful water use, such as depletion of aquifers (Kumar and Singh 2005).

- To environmentalists, importing virtual water goes against the idea of living sustainably within Nature's constraints. Why should so many people be living in places where there is not enough water to support them?

Ultimately, each of us must make our own decision as to how we feel about virtual water trade. What is clear is that virtual water trade is happening and will likely expand in the years ahead.

## 7.5. Desalination

Desalination—the removal of salts from saline or brackish water to produce freshwater—has generated a great deal of excitement, with its promise of a limitless, reliable supply of water derived from Earth's most abundant resource—salt water. Will desalination technology provide the deus ex machina that will solve our water-scarcity problems?

Early desalination technologies were dominated by *multistage flash distillation*, in which salt water is heated under vacuum to evaporate pure water, which is then condensed and captured. Newer desalination plants mostly use *reverse osmosis*, in which pure water is forced through membranes that reject salts. Both techniques are quite energy intensive and generate a concentrated brine ("concentrate") that must be disposed of. There is currently intense research and development in this field, focused both on improving current technologies (e.g., creating more efficient membranes that are less prone to fouling) and on developing new approaches (e.g., *forward osmosis*, in which removable solutes are added to the freshwater side to create an osmotic gradient that will drive water flow out of the salt water; McGinnis and Elimelech 2008).

Due to improvements in technology, desalination has made remarkable advances over the last several decades, whether measured by increases in installed capacity or by decreases in cost. While the global desalination capacity (~12 km³/yr; AQUASTAT 2009) represents only 0.3% of global water use, this capacity is growing at about 7% per year (Gleick et al. 2006). Desalination represents between 7% and 80% of water use for several energy-rich and water-scarce Middle Eastern countries (AQUASTAT 2009) that are essentially using the desalination process to convert oil into water. Other countries with significant desalination capacities are the US, Spain, and Japan, though the process represents a negligible fraction of water use in these countries (0.5%, 2.4%, and 0.5%, respectively). Approximately 20 new desalination facilities have been proposed for California, as the state struggles with drought, reduced water allocations from the Colorado River, and increasing water demand.

Costs for desalination have gone down dramatically over the last several decades, from around $9/m³ in the 1960s to generally less than $1/m³ today (Zhou and Tol 2005, Gleick et al. 2006). However, costs can vary significantly depending on the size of the project, the source of water, the technology used, the local costs of energy and land, and other factors. Some observers believe that costs will continue to come down with technological development, while others note that rising energy costs and approaching thermodynamic limits may lead to a leveling off of the cost curve.

Some projects have promised costs as low as ~$0.50/m$^3$, but actual costs have often been somewhat higher. The Tampa Bay project suffered from technical and institutional problems that delayed water production by 6 years and is now finally up and running (at 95,000 m$^3$/day) at a higher cost than anticipated: current costs are $0.87/m$^3$ (Allison Yount, Tampa Bay Water, personal communication), compared to the original projected cost of $0.55/m$^3$ (30-year average). The Ashkelon plant in Israel (330,000 m$^3$/day) is generally considered a success, but costs have risen from $0.54/m$^3$ (Gleick et al. 2006) to $0.66/m$^3$ (Gadi Rosenthal, personal communication). In contrast, the Singapore plant (137,000 m$^3$/day) was able to provide water for $0.48/m$^3$ during its first year of operation (2005–2006), compared to a preconstruction estimate of $0.78/m$^3$ (www.water-technology.net).

Reported estimates of costs can be misleading if they don't consider direct subsidies or the provision of subsidized energy or land. We should note also that desalination cost estimates often do not include the costs of delivering water to the consumer, which may be substantial, especially for cities that are at high elevations or are far from sources of salt water for desalination. Inland locations often focus on desalinating brackish groundwater, which has lower operation costs than seawater desalination (but higher costs for concentrate disposal).

It can be surprisingly difficult to make true cost comparisons between desalination and other water supply options. However, it is generally the case that desalination is still not cost-competitive with other sources—where those sources exist. We discuss water costs more in Chapter 12.

Desalination does have social and environmental impacts, although these are less significant than for dams and canals:

- Intakes: Like other water intakes in coastal areas, desalination plants can lead to mortality of fish and other organisms that are sucked into the intake pipes.
- Brine disposal: Desalination plants must dispose of a highly concentrated brine solution, usually by discharging into nearby coastal waters. This is generally not a problem in situations where the discharge is quickly dispersed and diluted to background salinities. However, more confined coastal ecosystems could be highly sensitive to elevated salinities and potential contaminants

in the discharge (e.g., chemicals used to clean equipment). Inland locations have few good options for brine disposal.

- Human health: Boron, which is found in seawater at concentrations of 4–7 mg/L, is not very efficiently removed by desalination, and some desalinated waters may exceed the WHO
drinking water guidelines of 0.5 mg/L. Many newer facilities
use specific processes to ensure adequate boron removal, which
can add significantly to the expense.

- Greenhouse gases: Because it is so energy-intensive, desalination
produces significant amounts of greenhouse gases that contribute to global climate change.

Philosophically, desalination is clearly part of a hard path approach.
Yet because it is still an emerging technology—and because its environmental and social impacts appear to be less than for large dam projects—
desalination facilities have aroused less opposition than dams. Still, several
projects have generated significant controversy, including the proposed
Poseidon plant in Carlsbad, California, which would be the largest seawater desalination plant in the US. In that case, the opposition seems to
stem mostly from three factors: the impacts on fish, the lack of attention
to conservation and limiting growth, and the involvement of a private,
for-profit company in providing water (Glaister 2004; Barringer 2009).

## 7.6. Wastewater Reclamation

In Frank Herbert's science fiction novel *Dune*, the human inhabitants of
the desert world Arrakis have developed ways to almost endlessly recycle
the water that they use, including capturing and drinking the water vapor
that they lose by sweating. While neither our water scarcity nor our technology have advanced to this point, there is increasing interest on Earth
in moving away from once-through systems and reclaiming previously
used water for reuse.

The reclamation field is full of jargon. Technically, *water reclamation* refers to a treatment process, which then allows *water reuse* by another user.
*Direct reuse* involves a piped connection between wastewater treatment
and the next user; *indirect reuse* involves release to the environment before
recapture by the next user. Additionally, reuse can be either *centralized* (involving effluent from a treatment plant) or *decentralized* (household-level
treatment and reuse); the latter is discussed in section 9.7, in the context
of household water conservation, while the former is the topic of this
section. *Water recycling* is the process of using water multiple times within
a given factory or other facility.

The greatest interest in the reclamation field has centered on tapping the stream of municipal wastewater. This is because agricultural water use produces relatively little return flow, while industrial effluents are difficult to reuse because they are highly variable in their quality and may include toxic contaminants. Municipal wastewater treatment is already in place for environmental protection in many countries, leading directly to the question: what additional treatment will it take before we can reuse this water, rather than discharging it as waste?

The answer, of course, depends on how we want to use the water. Current reuse generally falls into six categories: industrial, agricultural, groundwater recharge, urban nonpotable (e.g., landscape irrigation, firefighting), drinking, and environmental.[2] Each of these clearly has its own water quality requirements; indeed, some categories (e.g., irrigation) will have subcategories with different water quality needs (irrigation of crops that are consumed uncooked versus those that are processed before consumption). Water quality requirements also depend on whether the reuse is direct or indirect. When water is discharged to the environment before being reused (indirect reuse), it undergoes additional natural treatment, especially when it is released to groundwater and has a reasonably long travel time before being withdrawn for reuse.

For uses where human contact is likely, the main water quality concern is the transmission of disease through pathogenic organisms in the wastewater. We discuss sanitation and disease transmission in detail in Chapter 9, but here we simply note two basic points: use of poorly treated wastewater for irrigation or other uses can lead to transmission of disease; and given adequate treatment, wastewater can be safely reused even for food irrigation and drinking water.

According to AQUASTAT data, total world use of recycled wastewater amounts to about 22 km$^3$/yr (0.5% of world water use), of which the majority is in China. However, these data do not include the vast majority of wastewater reuse worldwide, which consists of the informal use of sewage (often poorly treated) on croplands in developing countries. It is estimated that perhaps 7% of the world's irrigated land receives wastewater (of which 90% is untreated), although this too may be an underestimate (Jiménez et al. 2010). The global health impacts of this type of activity are poorly understood but may be substantial.

---

2. The line between reusing wastewater for environmental purposes and simply discharging treated wastewater to a stream or lake is a fine one. The former category is used when the goal is a specific environmental enhancement (e.g., wetland creation, streamflow restoration) that would not take place without this water.

Israel has developed a sophisticated system for reclaiming and reusing wastewater safely. About 75% of Israel's municipal sewage undergoes secondary treatment, infiltration into soil, and recapture as groundwater for reuse in irrigation. Of the water used by the agricultural sector, more than half is reclaimed water. From the farmer's perspective, this situation has significant advantages: reclaimed wastewater is a reliable, nutrient-rich water source that is not sensitive to drought or other fluctuations, and is only expected to increase as population grows. At the same time, the long-term sustainability of this arrangement can be questioned, given the high concentrations of salts and trace metals in the reclaimed wastewater.

Potable use of reclaimed wastewater is rare, but both Singapore and Windhoek, Namibia, obtain a fraction of their drinking water needs from wastewater that has undergone extensive and highly monitored treatment. Although it may arouse some uneasy feelings, this type of planned reuse is certainly safer than the unregulated use of untreated wastewater in irrigation. It may also be superior to the unplanned, indirect reuse that is common along large rivers in the US, where downstream cities draw from river water that contains a significant fraction of treated sewage.[3]

## 7.7. Rainwater Harvesting

The same basic factor that drives dam-building—the need to overcome the temporal and spatial variability in rainfall—has also, for millennia, driven a variety of small-scale, decentralized approaches to capturing and storing rainwater. These diverse approaches, and the principles behind them, have now been united under the term *rainwater harvesting* (RWH). Techniques range from the traditional "tanks" of India (hundreds of thousands of small local reservoirs in valley bottoms), to the low embankments used by the Nabateans, Hopi, and others to direct runoff to their fields, or the rain barrels used by environmentally conscious suburban Americans. These approaches differ from dams primarily in scale, but also in that many of them capture rainwater before it becomes streamflow.

Rainwater harvesting has traditionally been most fully developed in regions with high variability in rainfall, such as desert and monsoonal climates, where—in the absence of human intervention—intense seasonal rains tend to run off quickly. Today, India, with its highly seasonal precipitation, provides perhaps the best example of a society attempting to recover old RWH techniques and develop new ones.

---

3. Connecticut is the only state in the US in which water quality standards forbid discharge of wastewater to a stream that is being used for drinking water.

RWH techniques in India (and elsewhere) can be classified by their purpose. In reality, these approaches are implemented in an astounding diversity of locally adapted ways in different parts of the country, so these classifications should be taken only as general descriptions:

- Household use: During the rainy season, runoff from rooftops or the ground surface is directed to underground storage tanks or to traditional circular holes lined with polished lime; these are used as drinking water sources throughout the year, either by individual households or by villages.
- Groundwater recharge: In urban areas, runoff from rooftops or the ground surface is directed into small infiltration basins. In rural areas, shallow ponds with pervious bottoms are created to capture rainwater and allow it to infiltrate into the soil. In both cases, the goal is to reverse the trend of falling water tables and allow more sustainable use of groundwater.
- Irrigation: Runoff is collected in excavated ponds (sometimes lined to prevent infiltration) and used for irrigation during the dry season. Alternatively, runoff farming is carried out, in which runoff is directed to fields and ponded there using small earthen dams and dikes. This technique can be used in arid areas to concentrate water onto a field from a relatively large area and can allow certain types of agriculture even in areas with remarkably low precipitation.

Nonprofit groups like the Centre for Science and Environment are working hard to promote the use of RWH as a solution to India's water-scarcity crisis. They see RWH as a sustainable alternative for India that can preclude the need for additional hard path solutions like more dams or the infamous "river interlinking" proposal, which aims to create a large number of interbasin links between areas of surplus and areas of deficit within the country.

Can RWH meet India's water needs? What would be the implications of widespread implementation of this decentralized technique? Not surprisingly, opinions differ widely on these questions. It seems important to note four basic points that should be part of the debate:

1. There are many examples of RWH success stories throughout India.
2. Scaling up RWH is likely to reduce streamflows in downstream areas, because RWH captures water that would otherwise produce surface water flow, at least in part.

3. The environmental and social costs of river interlinking are likely to be much larger than those of RWH.

4. India's population and economy today are much larger than in the bygone days of traditional water management. In the absence of dams, RWH alone could not provide water to a billion Indians.

## 7.8. Choosing Appropriate Technology

Which of the supply-side technologies discussed above should be implemented in order to address water scarcity? Alternatively, can demand-side approaches (conservation, efficiency) obviate the need for new supplies of any kind? The answers will clearly vary from situation to situation, and these are questions that we will continue to struggle with throughout this book.

We should make clear, however, that the technologies discussed above do not all serve the same purpose. Specifically, most of the approaches discussed in this chapter (water transport, virtual water trade, desalination, and wastewater reclamation) are aimed at *increasing the average amount of water available* in a region, whereas dams (both large and small) serve simply to *store water over time*, but do not increase average water availability.[4] Demand-side measures also fall into the first category, since they effectively increase the average water resource by enabling society to do more with less.

It is important, then, to differentiate between situations where water scarcity can be alleviated simply by increased storage (e.g., dams) versus situations that require changes in average supply or demand (e.g., virtual water trade). The former is more likely to occur where precipitation is highly variable (seasonally or interannually); the latter is more likely where cumulative human demand is approaching the limits of average water availability. Of course, the two factors interact: As the margin of safety narrows between average supply and average demand, even small fluctuations from the average can create temporary scarcity, requiring the construction of large amounts of storage to weather the temporary imbalance. Conversely, solutions that increase the margin of safety (e.g., virtual water trade) can alleviate the need for storage, since a country can weather a drought if its supply is large relative to its demand.

Brown and Lall (2006) analyze which countries require additional storage and which require changes to average supply and demand.[5] They

---

4. RWH falls somewhere in between: it is primarily a storage technique, but it can also increase blue water availability by shunting some green water to blue.

5. They refer to these as "hard" and "soft" water, respectively, noting that changes in average supply and demand should be accomplished through the soft methods of

find that a large group of countries from around the world fall into the latter category, while a smaller number (23) of mostly poor countries are in the former category of requiring more storage. They argue that this reflects underinvestment in large dam storage in developing countries and calculate how much additional storage should be built in each country. At the same time, they emphasize that for the larger group of countries, increased storage will not solve chronic scarcity problems.

Another key aspect of technological choice is the issue of scale. Large dams, water canals, and desalination plants are big, centrally managed projects, while rainwater harvesting and small dams are part of decentralized strategies involving a large number of small, individual efforts.[6] There are several important differences between these approaches. The large-scale projects tend to involve global expertise that is applied in a relatively uniform way from project to project, while the smaller-scale approach is usually tailored to local conditions and traditions. Each individual project at the larger scale requires massive financial investment (often by multinational corporations and international development banks), which can attract corruption but also can lead to greater national and international scrutiny. Smaller, locally managed technologies can benefit from community support but can also suffer from a lack of technical and management expertise. Assessment of effectiveness can be a problem for both large and small projects—the former because of the often-complex web of cascading positive and negative effects, and the latter because of the lack of hard data and the need to assess a large number of projects.

Which is better? Can the smaller-scale technologies cumulatively solve our water problems? There is no definitive answer to these questions, but maybe it is time to give the smaller-scale approach a try. As part of this effort, aid agencies, development banks, and governments must learn how to "let a thousand flowers bloom" while still rigorously assessing the effectiveness of their projects.

## 7.9. Conclusion

In this chapter, we examined questions of technological choice, investigating several water technologies in detail (dams, water transport, virtual water trade, wastewater reclamation, desalination, and rainwater harvesting)

---

policy reformation and conservation as well as virtual water trade. In my view, this terminology is confusing and ignores the array of choices for dealing with each type of problem.

6. Wastewater reclamation and conservation can operate at either scale, while virtual water trade is somewhere in between.

and starting to address the question of how to choose which technology is appropriate in a particular context. The technologies we discussed are very different in their scale, their social and environmental implications, and their ability to address different aspects of water problems.

Technological choice is in some sense both part of the water crisis and part of its solution. The technologies we have chosen over the last century or more have certainly contributed to many of the water problems we face today. Understanding these problems can help us understand how to make better choices and how to best utilize a mix of technologies to manage water more sustainably.

We turn now to a more detailed discussion of an aspect of the water crisis that has come up repeatedly in this chapter and previous ones: the damage to aquatic ecosystems (and human uses of those ecosystems) that has been a consequence—mostly unintended—of our water management. Here too we will be looking to understand the problems in order to understand how to move to a better balance between human water use and preservation of ecosystem health.

# 8

## Humans and Ecosystems: Finding the Right Balance

As noted in previous chapters, our management of water in the twentieth century involved spectacular feats of technology and engineering. We built huge numbers of massive dams in order to prevent flooding, capture water for our use, and produce electricity. We increased our withdrawals of water sixfold and used that water to expand our food production, our economy, and our standard of living. We built many, many miles of levees to protect ourselves from flooding. We drained wetlands to provide productive agricultural land. We built canals and straightened rivers for transporting our products through the continents. We increased our capacity to capture and grow fish in large rivers and lakes. All these actions have produced many benefits.

But our actions have also come with a huge cost: the degradation and even destruction of many aquatic ecosystems. Several formerly powerful rivers now run dry before reaching the sea. Other rivers flow to the sea, but their ancient seasonal fluctuations—so vital to their health—have been tamed and controlled. Dams block fish migration and break rivers up into disconnected segments. Pollution chokes our waterways and leads to depauperate aquatic communities and massive fish kills. Introduced species thrive in our altered waterways and drive out native species, lowering the biodiversity and resilience of our ecosystems. Urban streams are seriously sick—physically, chemically, and biologically. Many fisheries are collapsing from the combined effects of overharvesting, habitat degradation, and invasive species.

And this degradation of our rivers and lakes has not only affected "nature" but has caused great damage to society. Surface waters in many places are too contaminated to drink or even to swim in. Many rivers and lakes where we once caught abundant fish are now unproductive. The fish that we do catch are often contaminated with toxic chemicals. Dams have flooded productive lands and destroyed livelihoods and ways of life. Our deep spiritual and emotional connections to scenic waterscapes have been frayed. Floodplains that once supported a variety of productive activities are now cut off from the water and sediment that they need for nourishment. At the same time, damaging floods have increased despite—indeed, because of—our best efforts to tame rivers.

Is there a better way? Can we manage water in ways that balance human and ecosystem needs, or, to put it differently, that balance direct and indirect human uses? Can we act intelligently to restore aquatic ecosystems and the benefits that we derive from them?

## 8.1. Human Modifications to Aquatic Ecosystems

Table 8.1 summarizes the human activities that have altered freshwater ecosystems. Some of these activities fall squarely within the traditional scope of water management: water withdrawals, point source discharges, and dams. Other activities, such as land use change and the introduction of nonnative species, range a bit farther afield but nonetheless have clear impacts on the health of aquatic ecosystems. In any case, the numbers in Table 8.1 are impressive: we have made massive changes to water, land, and biota around the world.

Some have argued that these changes are truly *global* in their scope (e.g., Vorosmarty and Sahagian 2000). However, I believe that we need to be more precise in defining our terms. If by "global-scale alteration" we mean that these impacts are experienced almost everywhere in the world, then clearly the activities in Table 8.1 would qualify. But at the same time, the nature of these impacts is largely local or at most regional. With few exceptions, hydrological alteration and channelization and pollution in one watershed don't affect the health of adjacent watersheds, much less watersheds in other continents. One could argue that certain types of biotic impacts do extend beyond individual watersheds, such as overfishing of migratory species or the spread of invasive species from one watershed to another. However, even these impacts are really regional rather than global.

I would, however, note four ways that the activities in Table 8.1 have had consequences that have manifested (or could manifest) themselves

TABLE 8.1
Human activities that have had negative consequences
for aquatic ecosystems

| Action | Magnitude | Data Source | Impacts |
|---|---|---|---|
| water withdrawals[a] | ~4100 km³/yr | Shiklomanov (1999) | drying of rivers and lakes |
| effluent and return flow discharges[a] | ~1800 km³/yr | Shiklomanov (1999) | point and nonpoint source pollution, hydrologic alteration |
| large dams | ~48,000 | WCD (2000) | river fragmentation, hydrologic alteration, sediment trapping, etc. |
| small dams, other barriers | ~500,000? | extrapolation from US National Inventory of Dams | river fragmentation |
| reservoir capacity[a] | ~8300 km[c] | Chao et al. (2008) | increased water residence time, sedimentation, evaporation |
| physical impacts to rivers (channelization, etc)[b] | >500,000 km | Revenga et al. (2000) | habitat destruction, sediment disequilibrium |
| wetland loss | 480,000 km² (53% of original wetlands) in conterminous US alone[3] | Dahl (1990) | habitat destruction, changes in watershed flow pathways |
| land use change[d] | ~16,000,000 km² cropland ~34,000,000 km² pasture ~500,000 km² urban[e] | cropland and pasture: FAOSTAT (2007 data); urban: Potere and Schneider (2007) | changes in watershed flow pathways, nonpoint source pollution |
| fishing | 10.1 million tons/yr | FAO (2008) | depletion, species shifts, pollution, habitat destruction |
| aquaculture | 31.6 million tons/yr | FAO (2008) | pollution, displacement of native species, loss of genetic diversity |

TABLE 8.1 *continued*

| Action | Magnitude | Data Source | Impacts |
|---|---|---|---|
| introductions of nonnative species | >5600 introduced aquatic species recorded | www.fao.org/ fishery/dias/en | elimination of native species through competition and predation, global homogenization and loss of biodiversity |

a. For reference, global runoff (excluding Antarctica) is ~42,000 km³/yr; see Chapter 3.

b. Data are unavailable for the rest of the world. Wetland loss in the US was caused primarily by draining and filling, and had slowed dramatically by the 1980s.

c. Data are scarce for this category. The number shown is an estimate of "waterways altered for navigation" as of 1985.

d. For reference, the total area of Earth, excluding Antarctica, is ~135,000,000 km².

e. Depending on definitional and methodological issues, this number has been estimated to be as low as 280,000 km² or as high as 3,500,000 km².

at the global scale. The first is more of a curiosity than a serious concern: the anthropogenic redistribution of water on the planet has been large enough to change the distribution of mass over the globe and affect Earth's rotation on its axis. In particular, the large mass of water stored in reservoirs, mostly in northern temperate areas, is calculated to have increased the rate of Earth's rotation and led to a decrease in day length of about 8 microseconds (Chao 1995).

More important, the changing distribution of water has implications for the rate of sea level rise. This effect can operate in two ways. On the one hand, water that is stored in reservoirs, or that has seeped into the ground beneath reservoirs, is water that is not in the oceans; this results in a drop in sea level. On the other hand, activities like groundwater over-pumping or the drying of inland lakes and wetlands are taking water that was on land and ultimately moving it to the oceans, thus increasing the rate of sea level rise. With the large uncertainties around the magnitude of each of these effects, our best guess has been that they roughly cancel each other out. A new analysis by Chao et al. (2008) suggests that the sea level fall due to reservoir impoundment may be larger than previously calculated and has averaged about 0.55 mm yr⁻¹ for the last 50 years. This is significant relative to the total global sea level rise of about 2 mm yr⁻¹ and needs to be considered in trying to improve our understanding of the factors controlling the past and future rates of sea level rise.

The third type of global impact comes from a group of persistent organic pollutants (POPs). These compounds—including things like PCBs (polychlorinated biphenyls), DDT (dichlorodiphenyltrichloroethane), and dioxins—are so stable (persistent) that they last for many years in the environment. In addition, they have physical and chemical properties that allow them to be transported globally, primarily through evaporation followed by long-range atmospheric transport and redeposition. They have been found in aquatic ecosystems (and humans) throughout the world, even in places far removed from any local sources.

Finally, there is the possibility of large-scale changes in weather patterns due to changes in ET caused by land use change and consumptive water use. For example, conversion of forest to agricultural land in South Asia and sub-Saharan Africa could lead to changes in the Asian monsoon (Gordon et al. 2005), and deforestation in the Amazon could affect precipitation in the La Plata basin (Hoff 2009).

The remainder of this chapter focuses on local-scale impacts. We discuss the different types of impacts on aquatic ecosystems: hydrological (changes in water flow), physical (changes in sediment and the channel), chemical (water quality), and biological (direct and indirect impacts on organisms ranging from algae to fish). Note that we don't discuss the direct impacts of dams (e.g., river fragmentation), which were already covered in section 7.1. We then turn to examining the ways that the combination of these four factors leads to the degradation of streams in urban environments, and finally we look in a little more detail at the legal structure for protecting aquatic ecosystems in the US.

## 8.2. Hydrologic Alteration

Perhaps the most basic and most significant human impact on rivers and other aquatic ecosystems has been the alteration of natural hydrology. This has taken place primarily through the following activities:

- Water withdrawals: Withdrawing water from rivers, especially for consumptive use, leads to lower streamflows. Groundwater withdrawals can also affect streamflows, particularly in areas where groundwater and surface water are closely linked. Lakes and inland seas can also be impacted by withdrawals of water, either from the water body itself or from a river that feeds it (see Box 8.1).

- Return flows: Return flows from nonconsumptive uses can change the magnitude and timing of river flows. This is most dramatic for interbasin transfers, where the return flow is being discharged to a different watershed than it came from; this can lead to flows in the receiving watershed that are significantly higher than natural.

- Dams and reservoirs: Water that spends considerable time in reservoirs takes longer to reach the oceans. Vorosmarty et al. (1997) referred to this as "aging" of continental runoff and calculated that the water residence time in rivers has more than doubled globally, with an average increase of 31 days. Among the implications of this increased residence time are increased losses to evaporation.

- Dams and streamflow: In addition to their role in water withdrawals, dams often have significant impacts on the variability and timing of streamflows. Flood-control dams are designed to reduce the magnitude of high flows and may also serve to augment low flows, as water stored during floods is released during dry periods. Hydropower dams are often operated to provide peaking demand, which means that they will release more water during periods when there is higher demand for electricity; this may lead to massive fluctuations in streamflow on a daily basis.

- Land use change: As discussed in Chapter 6 (and in more detail in section 8.6, below), deforestation and urbanization can lead to flashier watersheds, in which stormflows are both higher and of shorter duration than in predevelopment conditions.

- Wetland drainage: Vast areas of wetlands have been drained ("reclaimed"), mostly for agricultural use. The artificial lowering of the water table leads to loss of this valuable habitat and its water storage capacity, which generally leads to faster water movement through watersheds. Additional, unintended consequences of wetland drainage can include land subsidence and acidic soil conditions.

Dramatic examples of hydrologic alteration are presented by large rivers that no longer reach the sea because of consumptive uses in the basin, such as the Colorado and the Huang (Yellow). This phenomenon is most likely to occur in the hot, dry season, when natural water availability is least and human withdrawals are typically greatest. As discussed in section 6.1, large rivers that go dry are the exception rather than the rule: late-twentieth-century trends in the annual discharge of most large rivers are influenced primarily by changes in precipitation, rather than human withdrawals (Milliman et al. 2008, Dai et al. 2009, Döll et al. 2009). No comprehensive analysis of small rivers is available, but they are probably

---

## Box 8.1. Desiccation of the Aral Sea

The poster child for the ecological costs of twentieth-century water management is the Aral Sea. This inland lake in Central Asia receives inflow from two large rivers, the Syr Darya and Amu Darya. Or it did, until Soviet central planners decided to use the water from those rivers to grow vast quantities of cotton.

The diversion of Aral Basin river water for agriculture started in earnest in the 1950s. The resulting decrease in river flow led to shrinkage of the lake, which now occupies about 10% of the area that it did in 1950. The lake has separated into two basins: the northern Small Aral and the southern Large Aral. Owing to the absence of freshwater inflow and the influence of evaporation, salinity has increased from about 10 psu to over 100 in the Large Aral (compared to 35 for seawater). Very few fish species can survive under these conditions, and the once-thriving fisheries of the Aral have collapsed. The recession of the lake shoreline has left former port cities far from the water and has exposed salt-laden and toxic sediments, which are blown around the region by windstorms. A number of chronic diseases have increased in the region, and life expectancy has dropped.

There are signs of hope for the Small Aral. Kazakhstan has built a dike to prevent outflow from the Small to the Large Aral, which has resulted in positive consequences for the smaller water body: an increase in water level, an increase in lake area, a decrease in salinity, and an apparent increase in fish catch. Improvements in irrigation efficiency could potentially increase the water flow from the Syr Darya and further improve conditions in the Small Aral.

---

more likely to experience periods of reduced or zero flow due to human withdrawals.

Even when the annual discharge of a river is unchanged, the river can still be highly impacted if the *distribution* of flow over time has been altered. Over the last decade or two, scientists have started to realize that keeping a river healthy means not just making sure that it doesn't dry up but also maintaining the flow pattern that makes up its particular natural "flow regime" (Poff et al. 1997).

## River Flow Regimes

Specifically, the flow regime is characterized by the *range* of different flows that a river experiences, and includes the *magnitude, timing, duration, frequency, rate of change,* and *predictability* of these flows. Different types of flow are thought to be important for maintaining different aspects of river health:

- Low flows serve to concentrate prey for aquatic predators and to provide dry enough conditions for certain floodplain plants to establish themselves. At the same time, low flows need to be high enough to ensure adequate habitat and connectivity within the channel and to prevent the temperature and dissolved oxygen from reaching levels that are toxic to the organisms of that river system.
- High flows that can move sediment are important for shaping the channel and for clearing fine sediment from spawning beds.
- The highest (flood) flows are necessary for connecting the floodplain to the channel. These flows allow access to floodplain food sources by aquatic organisms and provide river sediment and nutrients to the floodplain.
- The timing of different flows serves as a cue for fish migration and spawning and initiates lifecycle changes for aquatic insects.
- The rate of change of flows must be high enough to initiate natural biological changes, but not so unnaturally high that aquatic organisms are stranded by rapid drops in water level or washed out by rapid increases in flow.

Underlying the flow regime paradigm is the notion that streamflow serves as a "master variable," controlling the physical, chemical, and biological character of a given river. The biotic community in a particular river is adapted to that river's flow regime and to the conditions that are created by that flow regime. Changes to flow, then, will alter both the abiotic conditions and the biotic community, presumably to ones that are less desirable. The consequences of flow alteration can include accumulation of sediment in the channel, disruption of fish migration, encroachment of upland vegetation into the floodplain and the channel margins, impoverishment of the invertebrate community, invasion of nonnative species, loss of native biodiversity, and degradation of floodplain-dependent fisheries and agriculture (Box 8.2).

## Characterizing Hydrologic Alteration

How do we know how altered the flow regime is for a particular system? The best way to characterize hydrologic alteration is to compare a current hydrograph to a "natural" hydrograph. The natural hydrograph could

## Box 8.2. Dams and Hydrologic Alteration in Northern Nigeria

The Hadejia-Jama'are floodplain in northern Nigeria provides a good example of the downstream impacts of hydrologic alteration. This area, part of the Lake Chad basin, is located in a semiarid climate, on the southern border of the Sahel region. Yet it traditionally has experienced seasonal flooding (July–October) due to seasonal rains in the upstream (southern) portion of the watershed, combined with the very shallow slope and complex, braiding river channels of the floodplain area itself. This flooding has several beneficial effects, and a complex ecology of human use has grown up around it (Thompson and Polet 2000). This includes several types of agriculture over the course of the seasonal cycle: rain-fed cultivation of sorghum and millet in upland areas; flood cultivation of rice in flooded areas; irrigation of wheat and vegetables; and farming of a variety of crops in the floodplain after the rice harvest (at the end of the flood season) using the residual moisture ("flood recession agriculture"). Cattle are also grazed on the floodplains in the dry season, when the residual moisture from the floods allows natural vegetation growth in an otherwise arid region. In addition, the seasonal flooding allows fish to grow and breed rapidly throughout the flooded area, and different fishing activities have developed around this seasonal cycle. The flooding is also important for recharging groundwater, which can then be withdrawn during the dry season.

The Tiga Dam was constructed in 1974 in an upstream tributary of the Hadejia-Jama'are floodplain, primarily in order to provide water to the Kano River Irrigation Project. Two additional smaller dams were completed in 1992, and plans have been formulated to further expand irrigation and possibly construct an additional dam. The combination of the dams and the Sahelian drought has led to a dramatic decline in downstream flooding. The area inundated at the height of the flood season dropped from 2000–3000 $km^2$ in the 1960s to generally less than 1000 $km^2$ in the 1980s and 1990s. This led to a drop in fish and agricultural productivity, although farmers have adapted their practices to existing conditions (Thomas and Adams 1999). Barbier (2003) argues that when all floodplain uses are accounted for, the net value of water is much higher in the floodplain than in the upstream irrigation project. He estimates that full implementation of upstream irrigation plans would result in an annual net loss of ~\$20 million, while keeping to current levels of irrigation and implementing a regulated flood would restrict those losses to ~\$2–\$8 million. Dam removal would presumably be even better economically but is not feasible politically.

be a hydrograph for the same river for an earlier time period—before significant human intervention in the hydrology—but could also be the hydrograph for a pristine river that is thought to be otherwise very similar to the river in question. Examples of hydrographs that demonstrate clear patterns of hydrologic alteration are shown in Figure 8.1.

Some hydrologic changes, such as large decreases in seasonal flooding, can be identified by simple inspection of hydrographs. Others, such as smaller changes in timing or duration of flows, require more detailed examination of the hydrologic data. The Nature Conservancy has developed a comprehensive set of hydrologic indicators (Richter et al. 1996), along with software to extract values of these hydrologic indicators from flow datasets. This software, known as the Indicators of Hydrologic Alteration, is widely used to assess both the *degree* and the *type* of hydrologic alteration experienced by a given system.

Another approach for characterizing hydrologic alteration was recently proposed by Weiskel et al. (2007). The goal of this approach is not to characterize the detailed patterns of hydrologic change but to identify different types of human water use regimes and how they affect water balances in watersheds or aquifers. This involves estimating annual water budgets and using them to calculate two water use indicators: the *water use intensity* (WUI), which expresses what fraction of runoff in the watershed is withdrawn by people (range = 0–1); and the *human water balance* (HWB), which captures the extent to which there is net consumption or addition of water within a watershed (range = −1 [consumption] to +1 [addition]). Different values of these indicators define four patterns of impacts:

- Low WUI, near-zero HWB, "undeveloped"
- High WUI, near-zero HWB, "churned": most water in the river has been used by people, but withdrawals and discharges are in balance overall.
- High WUI, negative HWB, "depleted": consumptive use of water dominates.
- High WUI, positive HWB, "surcharged": interbasin transfers from other watersheds lead to higher-than-natural flows.

## Environmental Flow Requirements—Defining the Question

There has been increasing recognition that hydrologic alteration has serious harmful effects on the integrity of aquatic ecosystems. Likewise, we now recognize that *restoring flows* to degraded rivers can help restore the entire ecosystem (though there are often other issues that must be dealt

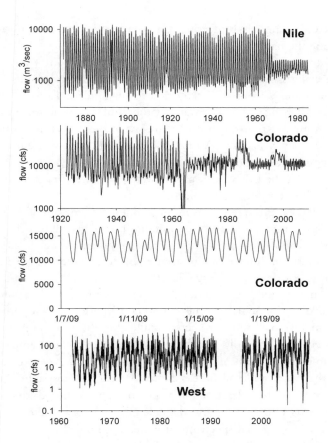

*Figure 8.1.* Examples of hydrologic alteration. Note different scales and units. Top: Nile River, Egypt, monthly data (source: www.rivdis.sr.unh.edu/download.html). The construction of the Aswan High Dam (1960–1970, reservoir began filling in 1964) led to a dramatic reduction in the annual flood and an increase in the annual low flows (as well as a shift in seasonal timing that can't be distinguished at this scale). Second from top: Colorado River at Lees Ferry, Arizona, monthly data (source: USGS 2009, station 09380000). The construction of the Glen Canyon Dam (1956–1966, reservoir began filling in 1963) led to similar effects to those seen on the Nile. (The construction of the Hoover Dam (1931–1936), downstream of Lees Ferry, did not affect flow at this site.) Third from top: Colorado River at Lees Ferry, Arizona, 15-minute data (source: USGS 2009, station 09380000). Production of hydropower for peaking electricity leads to very large changes in river flow on the time scale of several hours. Bottom: West River near Uxbridge, Massachusetts, daily data (source: USGS 2009, station 01111200). This small river has experienced little change in high and average flows but has seen a substantial reduction in the magnitude of the lowest low flows, as a result of human withdrawals.

with as well). Similarly, protecting relatively healthy streams must involve protecting the natural flow regime.

At the same time, human uses of water are vital and not easily given up, especially in an era of population growth and increasing scarcity. These uses are, of course, directly tied to manipulation of the flow regime, whether in the form of water withdrawals or in the form of water storage from wet to dry periods. Thus, while there is clearly a need to rebalance many rivers in favor of ecosystem protection, we are not simply free to completely restore natural flow regimes.

As a result, there has been great interest in answering the question, How much water does a river need?

I believe that this question is misguided, for two reasons. First, it tends to encourage a single answer—a minimum flow that will protect habitat. As discussed above, an environmental flow requirement needs to include not just one minimum flow but a range of flows: different baseflows in different seasons, small and large floods.

Second, this question assumes that scientists can provide a flow prescription that will achieve some hypothetical perfect balance in which ecosystem health is fully protected while the effect on human use is minimized. It is generally not the case, however, that flow impacts follow a simple threshold: "if we leave X cfs in the river, we can take the rest of the water and still have a perfectly healthy ecosystem." Instead, flow impacts are best described as a spectrum, from the healthiest, most freely flowing river to the most degraded, hydrologically altered system. As we move along that spectrum of increasing flow alteration, different ecosystem components will respond differently, with some experiencing gradual degradation and others experiencing threshold effects, each at a different level of hydrologic change. Ultimately, society—not scientists—must judge what level of degradation or restoration is acceptable for a particular river, given the other demands on that water.

The role of ecohydrologists, then, is to *describe the spectrum* as clearly as possible so that decision makers have the information they need. The question before us needs to be, How will different levels of alteration to the flow regime affect the structure and function of the aquatic ecosystem?

Even this question is quite a difficult one. Although scientists have successfully traced out some of the linkages between a particular flow level and specific river processes (both biotic and abiotic), these linkages are sometimes quite complex and hard to prove. More important, we simply don't have the resources to delineate these linkages for every river that we are concerned with.

## Environmental Flow Requirements—Methods

A large number of methods have been used to estimate environmental flow requirements, each of which deals somewhat differently with the challenges posed above. These methods are often grouped into the following four categories (Tharme 2003):

1. Hydrologic methods: These methods generally don't attempt to trace out specific linkages between hydrology and ecosystem health, but rather focus on maintaining or restoring some features of the river's natural hydrology—with the implicit assumption that fixing this "master variable" will protect or restore the ecosystem as a whole. The oldest and simplest hydrologic method is the Tennant Method, which sets a minimum flow as a percentage of the river's natural mean annual flow. A more recent example of a hydrologic method is Smakhtin et al.'s (2004) Environmental Water Requirement (EWR), which uses simple rules to calculate both low-flow and high-flow requirements for all large river basins globally; they calculate that these requirements constitute 20–50% of mean annual flows. A more sophisticated hydrologic method is The Nature Conservancy's Range of Variability Approach, which uses preimpact hydrologic data to calculate acceptable ranges for each of the 34 Indicators of Hydrologic Alteration parameters.

2. Hydraulic methods: This approach attempts to identify a level of flow where favorable hydraulic conditions are created. For example, in the Wetted Perimeter Method, measurements are made at several cross sections and used to estimate how the wetted surface area of the stream changes as a function of flow. These relationships can then be used to pick a flow level where a certain amount of surface area is available (e.g., where the entire bottom of a U-shaped channel is wetted).

3. Habitat simulation methods: These methods delineate how the available habitat—usually for a single fish species—changes as a function of flow. The most widely used of these methods is the Instream Flow Incremental Methodology (IFIM), which combines two components: measurements of how depth, velocity, and substrate change with flow; and a model that describes how depth, velocity, and substrate affect habitat suitability.

4. Holistic methods: Holistic approaches, which take into account multiple ecosystem components, were developed more recently,

mostly in South Africa and Australia. These tend to include not just minimum flows, but the entire flow regime, and often involve large interdisciplinary teams of experts studying a particular river in great detail.

Arthington et al. (2006) argue persuasively that we need to find the right middle ground between overly simplistic methods and overly labor-intensive ones. They claim that the simple hydrologic methods "have no documented empirical basis and the temptation to adopt them represents a grave risk to the future integrity and biodiversity of the world's riverine ecosystems." They further suggest that adoption of Smakhtin et al.'s (2004) EWR "would almost certainly cause profound ecological degradation." At the same time, they acknowledge that for most rivers where environmental flow requirements are needed, the resources are not available to convene expert panels and conduct detailed data gathering.

Arthington et al. (2006) propose an intermediate approach (more fully developed by Poff et al. 2010), in which streams within a region are classified into groups with reasonably similar natural hydrology. For each class of rivers, flow parameters are chosen (e.g., timing of peak flows) and the natural range for these parameters is identified. Then, flow-response relationships are established, which describe how different aspects of ecosystem health change when the values of the flow parameters depart from the natural range. The flow-response relationships may initially be based on theory or data from other sites, but over time, these relationships should be fine-tuned for each group of streams.

Regardless of the method used to calculate environmental flow requirements, actually implementing these recommendations can be a significant challenge, since it will generally involve substantial changes to water withdrawal, discharge, and storage.

## The Global Extent of Hydrologic Alteration

To what extent are rivers around the world currently suffering from degradation due to hydrologic alteration? The first attempt to quantitatively answer this question was Smakhtin et al.'s (2004) analysis, which identified many areas where human use is cutting into the EWR, including much of western North America, the MENA region, central and south Asia, and the Huang and Murray-Darling basins. Approximately 1.4 billion people were calculated to live in these "environmentally water scarce" basins.[1]

---

1. Smakhtin et al. (2004) define a Water Stress Indicator (WSI), which is identical

This was a very crude analysis, based on a simple comparison of annual withdrawals to calculated annual EWR.

More recently, Döll et al. (2009) used the WaterGAP model to prepare grid-based estimates of the degree of anthropogenic alteration to six hydrologic parameters, including average annual flows, low flows, seasonal amplitude and timing, and interannual variability. They found that the parameters with the greatest degree of alteration were (1) the low-flow parameter, Q90 (see section 3.1), which showed substantial decreases (i.e., lower low flows) for 26% of Earth's land area and substantial increases for 5%; and (2) the interannual variability parameter, which showed substantial increases (i.e., greater year-to-year variability) for 27% of the land area and substantial decreases for 8%.

## 8.3. Physical Alteration

As we discussed in section 2.4, river channels reflect the balance between sediment sources and water flows. In a healthy river, this balance results in a stable or slowly evolving channel and floodplain. Disruption to either sediment sources or water velocities can upset this equilibrium and lead to a channel that rapidly erodes or accumulates sediment.

Because of this linkage between water and sediment, the hydrologic changes noted in section 8.2 often lead to physical changes in rivers. Most commonly, an absence of high flows can lead to sediment accumulation in the channel and can deprive the floodplain of needed sediment.

In addition, humans have made many direct alterations to river channels, with a variety of goals: improved navigation, more efficient passage of floodwaters, land reclamation, creation of recreational opportunities, increased water storage, aesthetic improvement, and road crossings. These alterations typically involve straightening, widening, and deepening the channel or, in extreme cases, replacing it with concrete. Also common are the hardening of riverbanks through levees or riprap and the removal of snags (downed trees in the channel) and other coarse woody debris.

These changes can lead to a river system that is out of equilibrium, with negative consequences for both the ecosystem and people. For example, artificial channel straightening causes an increase in slope, which leads to higher velocities that can erode the stream banks and create an unstable channel. Artificial channel widening leads to lower velocities,

---

to the WTA indicator except that the environmental water requirement is subtracted from annual runoff when calculating availability. Environmentally water scarce basins are those where the WSI is greater than 1.

with associated sediment accumulation and a higher water level. Channel changes designed to pass floodwaters more quickly can lead to increased flooding downstream, as discussed in Chapter 4.

In addition, these highly engineered solutions can damage or destroy the habitat value of rivers. Natural streams are complex systems, with a variety of microhabitats available due to the presence of different water velocities, depths, and substrates. Hard engineering of river channels tends to flatten out these differences and replace them with a homogenized, sterile system. A concrete channel—with no sand beds for spawning, no woody debris to hide behind, no particulate organic matter to feed on, and no pools to rest in—is not a river. Further, if this concrete channel is designed to be large enough to carry extreme floods, it will be too wide for normal flow conditions, leading to shallow, hot, and poorly oxygenated water that very few fish can live in.

To what extent are physical alterations affecting rivers around the world? Global assessments of channel modifications are lacking, with the entry in Table 8.1 being an inadequate and dated estimate.

## 8.4. Chemical Degradation: Water Quality

Water quality is perhaps the issue where the links are strongest between ecosystem degradation and loss of human value. Highly polluted water bodies are both unhealthy for aquatic organisms and unsafe for human use, whether for drinking, swimming, or even agricultural and industrial uses.

Pollution issues are woven throughout this book. As discussed in Chapter 2, the most basic distinction in terms of pollutant sources is between the discrete *point sources* and the land use–based *nonpoint sources*. Each of these is discussed at different locations in the book:

- Water quality impacts of deforestation (a nonpoint source) are discussed in section 6.2.
- The water quality problems associated with dams are discussed in section 7.1.
- Urban runoff—conceptually a nonpoint source but often legally considered a point source—is discussed in section 8.6, below.
- Municipal point sources (sewage) are discussed in Chapter 9.
- Agricultural pollution—which can be either point or nonpoint pollution—is discussed in Chapter 10.
- Industrial point sources and the water quality impacts of resource extraction (forestry, mining) are discussed in Chapter 11.

What do we know about overall global water quality? Assessing the water quality of the world's freshwater resources in any kind of comprehensive and quantitative manner is a difficult task, one that we are still in the early stages of attempting. The UN's Global Environmental Monitoring System (GEMS Water) is relatively undeveloped and suffers from the fact that many countries either lack the resources to carry out systematic monitoring or are unwilling to share their data. An additional problem is the sheer complexity of water quality monitoring—which parameters to monitor? how frequently? at what cost? There are certainly many cases around the world of severe pollution of freshwater resources, but no comprehensive assessment exists.

## 8.5. Biotic Degradation

The hydrologic, physical, and chemical changes outlined above have produced freshwater habitats that are often dramatically different from native habitats, usually for the worse. This is often reflected in a degraded biotic community, whether because of poor water quality, improper substrate, lack of adequate seasonal flows, or the combination of many factors.

In addition to hydrological, chemical, and physical insults, native biota have to deal with two direct biotic impacts: overfishing and the introduction of nonnative species (Table 8.1). It is important to note that these impacts interact directly with other factors: nonnative species are more likely to outcompete natives where the habitat has been changed by modification of flow, substrate, or water quality (Kennard et al. 2005). Likewise, Poff et al. (2007) point out that dams and water withdrawals have led to homogenization of river flow regimes across the US; this loss of habitat diversity leads directly to a biotic homogenization and a loss of native biodiversity.

There are few overall assessments of the health of freshwater biota, in part due to the large number of different species that must be assessed. A study by The Nature Conservancy (Stein and Flack 1997) examined the conservation status of 13 different groups of plants and animals in the US and found that the four most endangered groups were all aquatic (freshwater mussels: 68% of species are at risk; crayfish: 51%; amphibians: 40%; freshwater fishes: 39%).

## 8.6. The Urban Stream Syndrome

Urbanization of a watershed leads to significant impacts on various components of stream ecosystems, including hydrology, geomorphology, water quality, and biology. In practice, these different impacts are difficult to tease apart and are really different aspects of what has been called the "urban stream syndrome" (Walsh et al. 2005)—the complex but consistent pattern of degradation of urban streams relative to their undeveloped (or agricultural) counterparts. This topic has been best studied in developed countries, where the effects of urbanization on aquatic ecosystems have generated much research and management activity. We begin this section by discussing the impervious surfaces that characterize urban watersheds, then proceed through the different types of impacts (hydrologic, geomorphic, chemical, and biotic), and close with a discussion of how to better manage urban streams.

## Impervious Surfaces

Most of the impacts of urbanization stem from the presence of *impervious surfaces*—areas where soil has been replaced or covered over with a surface that cannot absorb water, such as a building, road, parking lot, or driveway. Rainfall that hits impervious surfaces will run off right away to the nearest stream—or to the nearest storm drain, from where it will get transported in a pipe to the stream. This contrasts with the behavior of rainfall on a vegetated surface, where most of the water from a typical rain event gets absorbed into the soil and either infiltrates to groundwater (sustaining baseflow) or travels to the stream as stormflow—but relatively slowly, by subsurface flow in the soil.

Thus, impervious surfaces fundamentally change the behavior of water. For that reason, the amount of impervious cover in a watershed has been used as a master variable to predict and understand impacts on aquatic ecosystems.[2]

Table 8.2 shows the relationship between different land use categories and percent imperviousness, as measured in the Chesapeake Bay region by Cappiella and Brown (2001). This study found that ~55–75% of the impervious area in these land uses were "car habitat," that is, driveways, streets, and parking lots, rather than "people habitat," such as buildings.

Some studies have made the distinction between *total impervious area* (TIA) and *effective impervious area* (EIA). The latter, also called *directly connected impervious area*, refers to the impervious surface that connects

---

2. Impervious surface data for the US are available at 30-meter resolution from the Multi-Resolution Land Characteristics Consortium (www.mrlc.gov/). A map of imperviousness for the entire world at 1000 m resolution is available from the Digital Water Atlas (atlas.gwsp.org/index.php?option=com_content&task=view&id=196&Itemid=63).

TABLE 8.2
Relationship between different land uses and impervious cover
(data from Cappiella and Brown 2001)

| Land Use | Percent Imperviousness (mean and standard error) |
|---|---|
| agriculture | 1.9 ± 0.3 |
| open urban land | 8.6 ± 1.6 |
| residential: 2 acre lots | 10.6 ± 0.6 |
| residential: 1 acre lots | 14.3 ± 0.5 |
| residential: 1/2 acre lots | 21.2 ± 0.8 |
| residential: 1/4 acre lots | 27.8 ± 0.6 |
| residential: 1/8 acre lots | 32.6 ± 1.6 |
| residential: townhouses | 40.9 ± 1.4 |
| residential: apartments | 44.4 ± 2.0 |
| institutional | 34.4 ± 3.4 |
| light industrial | 53.4 ± 2.8 |
| commercial | 72.2 ± 2.0 |

directly to a water body through surface runoff or the storm sewer network. Thus, for example, a house whose gutter system directs water to a lawn would not be considered part of the EIA, since that water will have the chance to infiltrate into soil. In contrast, EIA would include a house where stormwater is directed to a driveway, from which it flows to the street, then into a catch basin, and then to a stream. While EIA is probably a better measure of the likelihood of flooding and other impacts of imperviousness, it is a difficult parameter to estimate, and TIA is more widely used.

What is the nature of the relationship between imperviousness and stream degradation? Three models have been suggested (Figure 8.2): (A) a linear decline in stream health with increasing imperviousness; (B) a linear decline that is complete at an intermediate level of imperviousness, meaning that streams with intermediate and high TIA are equally likely to be highly degraded; and (C) a linear decline beyond a certain threshold, meaning that stream health can be maintained up to a certain level of imperviousness. There is evidence for each of these models in different situations, depending on both the location and the parameter being used to express stream health (Walsh et al. 2005). The Center for Watershed Protection and other watershed groups often cite 10% TIA as a threshold value beyond which significant degradation can be expected unless serious efforts are made to mitigate the effects of imperviousness.

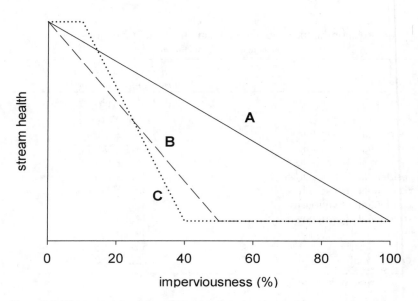

*Figure 8.2*. Three models of the relationship between imperviousness and stream health. Redrawn from Walsh et al. (2005).

It is important to note, however, that each watershed is different and that the relationships between TIA and degradation inevitably show a high degree of scatter. This is in part because other factors besides TIA can affect the response of urban watersheds; these factors have arguably been understudied because of the emphasis on impervious surfaces. Some of these factors include the compaction of soils in the "pervious" areas of the watershed; the *locations* of the impervious areas within the watershed; the size of the watershed; the nature of the storm sewer network; and other basic watershed features that modify the response of the watershed to imperviousness (e.g., slope, geology, soils). The total area of wetlands and ponds that have been filled and replaced with impervious surface may be a particularly important parameter to look at, since these ecosystems naturally slow down runoff.

## Hydrology

It has been understood for many years that urbanization leads to an increase in flooding. Urban streams are uniformly "flashier" than their undeveloped counterparts, with a quicker and higher stormflow peak and a more rapid return to baseflow. Leopold (1968) showed that the magnitude of the mean annual flood increases dramatically as imperviousness and storm sewer coverage rise. More recently, Booth and Jackson (1997) found that in watersheds with about 10% TIA, the streamflow that used to represent the 10-year flood now occurs much more frequently—with a 2-year recurrence interval. At levels of TIA above 10%, the 2-year flood is actually larger than the former 10-year flood.

One might expect that urban streams would also have lower baseflow than undeveloped streams, since the rapid runoff of precipitation leads to less recharge of the groundwater that sustains baseflow. While this effect has been observed in some cases, it is not a consistent feature of urban streams, perhaps because of other sources of water to sustain baseflow in urban areas, such as leaking water supply infrastructure.

## Geomorphology

The increased frequency of high flows in urban watersheds can lead to deepening and/or widening of the stream channel, as the high water velocities scour the stream bed and banks. For example, Booth and Jackson (1997) found that stream channels experienced erosion and instability when their watersheds exceeded 10% TIA. At the same time, poorly controlled land clearing in urbanizing watersheds can lead to abnormally high erosion in the watershed, with resulting sediment deposition in streams. This sediment can smother the gravel habitat needed by some fish for spawning.

The structural instability of urban streams commonly leads to direct human efforts to engineer the channel for greater stability and more efficient floodwater passage. In addition, the high property values, frequent road crossings, and high population density of urban areas often set in motion other channel and floodplain modifications of the types discussed in section 8.3.

## Water Quality: Urban Runoff

Wet-weather flows from urban areas tend to pick up pollutants from the land surface and carry them to the nearest stream. These pollutants typically include petroleum products from roads and parking lots; salt from winter road treatment; nutrients from lawn fertilizers, septic systems, pet waste, and atmospheric deposition; metals from roofs, car wear, and other sources; and bacteria from pet waste and septic systems. Note that urbanization increases pollutant loads in two ways: by increasing sources of pollution in the landscape and by shifting the pathways of water flow, which decreases the opportunities for pollutants to be removed as water moves through soils.

The levels of pollutants in urban stormwater can be quite variable, depending on the exact land use, the type of storm event, and other variables. The Nationwide Urban Runoff Program (NURP) in the US (1979–1983) was a concerted effort to obtain and compile data on the water quality of urban stormwater. Results showed that stormwater generated by high-intensity urban land uses was generally quite polluted and for some pollutants (such as lead and suspended sediments) could even be comparable to raw sewage.

The polluted runoff from urban areas is often thought of as nonpoint source pollution, since it originates from diffuse sources over a large area, rather than from a single location, as in the case of a municipal wastewater treatment plant, for example. Technically, however, under the Clean Water Act, urban runoff is regulated as a point source, since it typically flows through storm sewer pipes and discharges to streams from distinct outfalls. As a result, many municipalities are now required to obtain stormwater NPDES permits for their stormwater sewers (referred to as Municipal Separate Storm Sewer Systems or MS4s).

In addition to stormwater, many urban streams have other water quality issues, including polluted groundwater flow, industrial and municipal discharges, legacy pollutants (sediments contaminated with long-lived contaminants from the unregulated pollution of previous generations), and combined sewer overflows.

## Water Quality: Combined Sewer Overflows (CSOs)

Some communities have *combined sewers* that are designed to carry both the sewage from people's homes and the stormwater that enters from streets, roofs, and the like. The advantage of this approach is that stormwater is treated at the sewage treatment plant before being released to rivers. The disadvantage is that during large rain events, the sewer may not be able to carry the combined volume of rainwater and sewage. In this case, there is generally an overflow pipe that allows the excess to be discharged to a nearby stream. When this happens, the effluent that is released includes raw (untreated) sewage mixed with stormwater.

Dealing with CSOs has proven to be an expensive and vexing problem. One solution—separating the combined sewers into storm and sanitary sewers—has been implemented in some places but has several disadvantages:

- It involves extensive digging up of underground sewer lines, which is very expensive and time consuming.

- It creates a flow of urban stormwater to streams even during small rain events that formerly received treatment at the STP.
- It sometimes does not completely solve the overflow problem, in particular when there are leaky pipes that allow water infiltration during storms; these overflows are then referred to as sanitary sewer overflows (SSOs).

As a result, other options have been tried, including constructing underground tanks for storing combined sewage until the storm has passed and then pumping the combined sewage to the STP.

## Biotic Impacts

Given the hydrologic, geomorphic, and water quality impacts discussed above, it should come as no surprise that urban streams do not provide good habitat for aquatic organisms. In extreme cases, such as underground channels or streams receiving toxic effluents, urban streams may be practically devoid of life. More commonly, these streams are dominated by organisms that are relatively tolerant of pollution and degraded habitat.

In terms of ecological processes, urban streams are impoverished as well. Their ability to take up nutrients and retain organic matter has been shown to be diminished. Not enough research has been done for several other ecological parameters (Walsh et al. 2005).

## Managing the Urban Stream Syndrome

What can we do to better manage urban watersheds and restore urban streams? Given the numerous insults to these systems, improving their health requires a variety of different approaches, targeted at different scales, ranging from the stream itself to the larger watershed.

Stream. In-stream restoration can be used to directly improve stream stability and habitat quality. This can include dam removal, "daylighting" of buried rivers, replacement of streambank riprap with vegetation, and improvement of in-stream habitat (e.g., by adding woody debris). These measures require a deep understanding of how rivers function geomorphically and biologically. Restoration of physically degraded rivers needs to intelligently increase channel complexity, adjust channel capacity to new hydrologic regimes, and work with, rather than against, the power of water. In-stream restoration might also include improvement of water quality by removal or mitigation of pollutant sources such as CSOs, point source discharges, and contaminated sediments. In-stream restoration is unlikely to be fully successful if attention is not paid to floodplain and watershed processes as well.

Riparian zone. The riparian zone (the area immediately adjacent to the stream) should be a particular target for protection and restoration. When left as an intact vegetated system, this part of the landscape is quite valuable in providing room for the river to flood, slowing down stormwater en route to the stream, and removing pollutants through biological and physical processes. These last two functions of the riparian zone will be undermined if storm sewers simply carry water under the riparian zone directly to the stream.

Watershed. Successful urban stream restoration must address the watershed and its role in generating stormwater. One approach involves *structural stormwater best management practices* (BMPs): structures that are designed to slow down the movement of water to the stream, and in some cases, also provide water quality improvement. These include several categories:

- *Infiltration systems* are trenches or depressions where water can infiltrate into the ground rather than quickly running off.
- *Detention basins*, also referred to as dry basins, are depressions where water is temporarily ponded during a storm event and slowly released. This slows water down but does not provide much water quality benefit, since the residence time of water in the basin is relatively short.
- *Retention ponds*, or wet ponds, are similar to detention basins, except that some volume of water is stored even between storms. This increases the residence time and allows settling of sediments (and their associated contaminants) and uptake of nutrients.
- *Constructed wetlands* can provide significant water quality improvement, since the presence of plants and the dominance of subsurface flow can lead to efficient uptake of nutrients and other pollutants.
- *End-of-pipe systems* are proprietary vendor-supplied technologies that can be placed into storm sewers and are generally designed to remove coarse sediments (which sink to the bottom) and oils and floatables (which rise to the top).

In addition, *nonstructural BMPs* can be used to improve watershed stormwater management. Nonstructural BMPs include educating homeowners on how to prevent pollution and retain stormwater on-site, detecting and eliminating illegal connections of household sewage to storm sewer networks, and street-cleaning to remove potential pollution sources. These activities recognize that the stormwater problem, due to its diffuse "nonpoint" nature, is one in which homeowners and citizens must play an important role.

An emerging approach to urban watershed management is _low-impact site design_, which aims to design new construction in ways that minimize EIA and maximize stormwater retention in the soil. Some of the practices that can be involved include clustered housing and narrower roadways to minimize road surfaces; pervious paving, which allows water to infiltrate; taller houses and shared driveways to minimize imperviousness within each lot; and installation of rain gardens and vegetated swales to capture and use rainwater.

Any one of these types of activities in isolation is likely to have relatively little impact. But the combination of them all can do much toward re-creating a more natural hydrology, restoring the integrity of the river channel, improving water and habitat quality, and making the urban stream a more valuable resource for aquatic organisms and people alike.

## 8.7. Protecting Aquatic Ecosystems: The Clean Water Act and the Endangered Species Act

Given all the potential threats to rivers and other aquatic ecosystems, how do we assess how degraded a given ecosystem is? How do we determine what level of degradation is acceptable? And when this level is exceeded, how do we ensure that protection and restoration will occur? This section explores the way that these questions are answered in the US by the Clean Water Act (CWA) and, to a lesser extent, the Endangered Species Act (ESA).

### CWA Water Quality Standards and Assessment

As noted in section 2.3, the CWA aims to protect the "physical, chemical, and biological integrity" of aquatic ecosystems. In practice, the CWA defines integrity as the state of meeting all relevant water quality criteria. In particular, Section 303 of the CWA calls for states to establish _ambient water quality standards_, comprised of four components:

1. _Designated uses_ for each water body (e.g., swimming, aquatic habitat)
2. _Water quality criteria_ (WQC) describing levels of different parameters that are sufficient to protect each designated use; water

quality criteria usually include acceptable levels of chemical pollutants and descriptions of water body condition and target biotic communities

3. *An anti-degradation policy*, designed to prevent healthy water bodies from degrading to conditions that just barely meet water quality criteria

4. A general description of how the water quality standards will be implemented

WQC are meant to reflect the basic goals of the CWA that our waters be "fishable and swimmable," but the criteria are allowed to vary depending on the designated use. Thus, for example, in Connecticut, a Class B water (designated uses: "habitat for fish and other aquatic life and wildlife; recreation; navigation; and industrial and agricultural water supply") may have aesthetics that are "good to excellent," while a Class A water (designated uses: all of the above plus potential drinking water supplies) must have aesthetics that are "uniformly excellent." At the same time, many criteria are the same for both types of waters (e.g., DO at least 5 mg/L).

For the toxic pollutants (e.g., metals), separate criteria are calculated for protecting humans and aquatic organisms. The aquatic organism criteria include both acute criteria (levels that should not be exceeded even for a one-hour time period) and chronic criteria (somewhat lower levels that can be safely exceeded for a short time but should not be exceeded for a 4-day time period). Setting toxicity-based criteria requires a great deal of toxicity testing data and can be quite controversial, but the EPA has extensive guidance on how to carry out this process and has calculated numeric criteria for more than 150 pollutants.

Given a set of criteria that apply to a given water body, how does one evaluate whether these are being met? Under Section 305(b) of the CWA, the states must assess whether water quality criteria are being met; the EPA then compiles the state assessments into a biennial report to Congress. It is important to note that 305(b) reports from different states are not directly comparable, because the states can differ in their choices on three important questions:

1. Which parameters are included in the WQC and in the assessment? All states include the basic conventional and toxic pollutants in their WQC and 305(b) reports, but there is much less consistency when it comes to parameters that are not purely chemical, such as the level of in-stream flow, the degree of channel alteration, or the health of plants, invertebrates, and fish. These

issues are certainly part of "integrity" and should be included.

2. How frequently is a given water body monitored? Ecosystem health—especially water quality—can be quite variable over time, especially in rivers, where changes can occur over several different time scales (diurnal, storm, seasonal, long-term). Yet resource constraints necessitate limited sampling—often only four to eight times per year. Sampling programs should, at a minimum, try to capture occasional storm events, since water quality during storms may be quite different than during baseflow (Traister and Anisfeld 2006).

3. What is the spatial intensity and geographic distribution of monitoring? State sampling programs have often focused on larger or more accessible water bodies, which are not necessarily a representative sample of all water bodies in the state. The EPA has been pushing the states to use a more statistically valid approach to sampling, although it is important at the same time to continue data collection at long-term sites in order to detect trends over time.

One assessment tool that has become increasingly popular is *biomonitoring*, in which a biotic component of the aquatic ecosystem is monitored directly, instead of, or in addition to, traditional chemical water quality monitoring. Given that biotic health is a key component of ecosystem integrity and should be included in WQC, this makes perfect sense. The most commonly used bioindicators are *periphyton* (riverbed algae), *benthic macroinvertebrates* (insects and other small animals that live on the river bottom), and fish. For each group, the number and types of organisms found reveal something important about the health of the ecosystem as a whole.

The advantages of biomonitoring include the following (Karr and Chu 1999):

- Biotic health is in itself a key aspect of ecosystem health.
- The health of biota reflects conditions in the water body over a relatively long time period, so the issues of short-term variability in water quality can be sidestepped.
- Biomonitoring is relatively quick and cheap to perform, so more sites can be monitored.
- Biomonitoring tends to be more understandable and attractive to laypeople, allowing both greater public interest in the results and greater volunteer participation in monitoring.

## Impairments and TMDLs

Section 303(d) of the CWA requires states to identify water bodies that are "impaired," that is, not meeting the WQC that correspond to their designated uses. Of particular concern are those water bodies where impairment exists despite the implementation of technology-based effluent standards for point sources (see section 2.3). This may happen either because there are multiple point sources discharging to a relatively small water body or because nonpoint sources are contributing large amounts of pollutants.

In these cases, the state (or EPA) is required to prepare a Total Maximum Daily Load (TMDL), which is a determination of how much pollution this water body can receive, along with an allocation of that pollution among the different sources. The first part of this—the determination of the maximum pollution load—is conceptually fairly simple, though its scientific basis is somewhat questionable. It involves back-calculating the maximum acceptable load based on the water quality criterion and the dilution provided by the water body. The second part—the allocation of loads—can be politically complicated, often evoking finger-pointing and battles among the different point and nonpoint sources. Part of the complication is the difficulty of actually measuring the diffuse flux of pollutants from nonpoint sources or implementing measures that will reliably and measurably reduce that flux.

The TMDL approach is often held up as part of a holistic, watershed-based approach to ecosystem restoration. Indeed, the strength of this approach lies in two important features: its use of ecosystem health (as expressed by WQC) as an endpoint (at least in theory) and its call to evaluate and control *all* sources of pollution. This contrasts with the technology-based effluent limits that have dominated water quality control since the 1970s, which have been uniformly applied regardless of the health of the receiving water and which have controlled only point sources. Of course, the strength of the technology-based approach is precisely its simplicity and uniformity. It is no surprise that the technology-based approach was implemented relatively quickly after passage of the CWA, while the TMDL approach has really come into play only in the last decade and continues to move at a slow pace.

An example of an approved TMDL is discussed in Box 8.3.

## Box 8.3. TMDL for Eagleville Brook, Connecticut

An innovative TMDL (CT DEP 2007) was recently developed for Eagleville Brook, a small (6.2 km²) watershed in Mansfield, Connecticut. The upper portion of the watershed is relatively urbanized (TIA as high as 27%), in large part due to the presence of a campus of the University of Connecticut (UConn). Biomonitoring of the stream has revealed that the macroinvertebrate and fish communities are degraded, resulting in an impairment designation, but without revealing the precise nature of the stressor. The Connecticut Department of Environmental Protection (CT DEP) concluded that the cause of the impairment was "a complex array of pollutants transported by stormwater," where the term *pollutant* here is meant to include hydrological and physical issues as well as water quality—the whole collection of problems described by the "urban stream syndrome."

How do you write a TMDL for a complex array of pollutants? CT DEP chose to use percent impervious cover as a surrogate for stormwater impacts, due to the high correlation between watershed imperviousness and stream degradation. The threshold for impairment was set at 11% impervious cover (plus a 1% margin of safety before real impacts are expected to begin at 12% TIA). The TMDL was thus written in terms of how much of a reduction would be needed in effective impervious cover, most likely the first time that such a TMDL has been created.

This approach seems to make a great deal of sense. The basic problem at Eagleville Brook is a land use one, and the impervious cover approach taken in this TMDL tackles that issue head-on. At the same time, it is not clear how effective the TMDL will be in generating activities that actually reduce effective imperviousness. The TMDL refers to four such activities: reducing impervious cover where practical (e.g., by replacing paved parking lots with pervious ones); disconnecting impervious cover (e.g., by redirecting roof leaders); minimizing additional imperviousness, especially along stream buffers; and installing engineered stormwater BMPs to mitigate the effects of current imperviousness. But the TMDL does not give DEP any additional authority to require these actions, which are primarily under the control of UConn and the town of Mansfield. Nonetheless, the TMDL does bring attention to the problem and provides a framework for increased funding and motivation.

## The Status of America's Waters

Since 2001, the EPA has been encouraging the states to submit "integrated reports" that include both the 305(b) report (assessment of the state's waters) and the 303(d) list (list of waters requiring a TMDL). This information is now readily available online and can be used by the public in a variety of ways, such as understanding the health of a particular water body or finding all impaired water bodies in a given state.

In examining the integrated reports database, one should be aware of these points:

- Assessed water bodies can be listed as "good" (all WQC met), "impaired" (some WQC not met), or "threatened" (all WQC met, but water quality is deteriorating toward impairment).
- Not all impaired water bodies require TMDLs. In particular, a TMDL is not required if existing pollution controls are likely to remove the impairment or if the impairment is due not to a "pollutant" but to other forms of "pollution," such as hydrologic modification or changes to the channel.
- Water bodies are listed by "segment," where each segment is meant to be a relatively homogeneous part of the water body.
- The integrated report includes information on eight categories of waters: rivers and streams; lakes, reservoirs, and ponds; bays and estuaries; coastal shoreline; ocean and near coastal; wetlands; Great Lakes shoreline; and Great Lakes open water. Only a fraction of the area of each category is assessed in a given year.
- Because of all the differences noted above in how states carry out assessments (including which water bodies are examined), the integrated reports database is not a good tool for examining trends over time or even for comparisons among different states. Still, it provides a rough snapshot of how the US is doing at meeting the goals of the CWA.

The most recent data[3] for rivers and streams indicate that, of the assessed waters (which represent 26% of total river kilometers), about 50% are impaired, 50% are rated good, and a small fraction are rated as threatened. The main causes and sources of impairment to rivers nationwide are listed in Tables 8.3 and 8.4, respectively.

---

3. This is mostly based on reports from 2008, though for some states, the data are from previous years (as early as 2002).

TABLE 8.3
Top causes of impairments in US rivers and streams

| Cause | River Kilometers Impaired |
|---|---|
| pathogens | 227,700 |
| sediment | 172,800 |
| organic enrichment/oxygen depletion | 134,100 |
| habitat alterations | 132,600 |
| PCBs | 117,600 |
| metals (other than mercury) | 103,500 |
| flow alterations | 90,700 |
| temperature | 75,200 |
| cause unknown | 56,900 |
| salinity | 50,900 |
| other causes | 282,200 |

TABLE 8.4
Top sources of impairments in US rivers and streams

| Source | River Kilometers Impaired |
|---|---|
| agriculture | 166,700 |
| unknown | 124,900 |
| atmospheric deposition | 103,100 |
| hydromodification | 96,400 |
| natural/wildlife | 86,000 |
| unspecified nonpoint source | 71,300 |
| municipal discharge/sewage | 61,100 |
| habitat alterations (not directly related to hydromodification) | 53,100 |
| urban-related runoff/stormwater | 51,600 |
| resource extraction | 40,900 |
| other sources | 149,400 |

## Endangered Species Act

The Endangered Species Act (ESA), passed in 1973, has played a large role in protecting aquatic ecosystems in the US. The goal of the ESA is to protect species that are in danger of extinction, specifically those formally *listed* by the Fish and Wildlife Service (FWS) or National Marine Fisheries Service (NMFS, also known as NOAA Fisheries) as "endangered" or "threatened." Section 7 of the ESA prohibits all federal agencies from funding or authorizing any actions that will jeopardize listed species or their "critical habitat." Since almost any management of a reasonably sized river involves federal permitting or funding, the ESA provides a very strong tool for protecting an entire riverine ecosystem as habitat for a listed species. As a result, the ESA has been used by environmentalists to stop dams from being built, to prescribe flow releases, to prevent water withdrawals, to force improved water quality, and even to require dam removal.

The use of the ESA in this way has proven controversial. Some argue that those bringing ESA lawsuits don't really care about the endangered species but rather are using the ESA as a tool to rebalance water management toward broader environmental goals. In response, many environmentalists are happy to concede the point that their goals are broader than simply protection of one species, but they point out that using endangered species as a "canary in the coal mine"—a warning of environmental degradation—is consistent with the logic of the ESA.

The mechanics of implementing Section 7 can be quite complex. The *action agency*—the federal agency undertaking the action that might jeopardize critical habitat—must consult with either FWS (for freshwater species) or NMFS (for marine species). If warranted, the science agency prepares a *biological opinion* (BiOp), which includes its determination of whether the proposed action would jeopardize the listed species, along with any mitigation measures or alternative actions that it sees as advisable. While a BiOp is, in theory, based solely on the best available science, it is often the case that the science to support a decision is incomplete and open to interpretation, and so the process can sometimes be subject to considerable political pressure.

## 8.8. Conclusion

In this chapter, we discussed in some detail the various types of human impacts on aquatic ecosystems, especially rivers: changes in the amount of water that they receive; changes to the basic shape of the channel; degradation of the quality of the water; and disruption of the organisms that live there. These effects are all interlinked in a variety of ways, and the most human-impacted systems typically suffer from all of them.

Humans have always altered aquatic ecosystems, but the scale of these impacts has expanded dramatically in recent centuries. Fortunately, awareness of the issue has also grown, especially over the last 50 years, and the desire to protect and restore rivers, lakes, and coastal waters has been a key part of the environmental movement. As a result, we have made substantial progress, especially in developed countries like the US, in combating pollution and avoiding new dams. But there is still much to do.

Hydrologic and geomorphic impacts often get less attention than they should, and even on the water quality front, there are big gaps in protection. Half the assessed rivers in the US are impaired, and while TMDLs will be written for many of them (at great expense), it is not clear that we really have the tools—or the political will—to restore them to health.

As noted at the beginning of this chapter, when ecosystems are degraded, it is not just fish that suffer but people as well. We turn now to the issue for which this connection is felt most strongly—drinking water and other household uses.

# 9

# Overconsumption and Underconsumption: Water for Households and Health

Only a small fraction of the water withdrawn worldwide goes to households for direct uses, but these uses are the most basic and essential of all. In many developing countries, lack of access to safe water and sanitation services has devastating health consequences. Providing clean water to these households is a moral imperative of the highest order. At the same time, households in rich countries have access to water supply and sanitation systems that are much safer, but still have their problems. In addition, the ready availability of inexpensive, high-quality water in developed countries often leads to wasteful, excessive use, contributing to significant environmental and social problems. Reducing these household withdrawals to a reasonable level can free up water for other vital needs, including sustaining aquatic ecosystems.

We begin the chapter by outlining in general terms the links between water and health. This is followed by a detailed discussion of the state of water and sanitation in both developed and developing countries and how they might be improved. We then examine the natural groundwater contaminants arsenic and fluoride and look at the rise of the bottled water industry. We close the chapter by analyzing the potential for water conservation in household water use.

## 9.1. Water and Health

Water carries a dual status as both absolute necessity for health and potential bearer of disease.

To live a healthy life, people need a minimum amount of water for drinking. In addition, we need water for preparing and cooking food and for cleaning ourselves, our homes, and our clothes. Gleick (2000) has estimated a minimum requirement of 50 liters per person per day (Lpcd) of clean water for these uses; others put the number lower, at perhaps 20 Lpcd.

This same life-giving substance can also bring danger and death. Water-related health issues can be broken down into several categories:

- drowning
- burden of carrying: Water is heavy, and there can be health effects associated with regularly hauling large quantities of water from faraway sources.
- chemical and radiological contamination: Exposure to toxic chemicals or radionuclides in drinking or bathing water can lead to cancer as well as other acute and chronic symptoms.
- biological contamination: A large number of water-related pathogens—including various bacteria, viruses, helminths (parasitic worms), and protozoa—can cause infection, disease, and death (Box 9.1).

How do we protect ourselves from these risks? The answer involves six overlapping tools: source protection, water treatment, testing, water distribution, sanitation, and sewage treatment. *Source protection* means keeping water sources as free as possible from biological and chemical contamination, including protecting lands around water sources from hazardous activities. *Treatment* of water before use, either at a centralized location or at the household level, can reduce or eliminate many contaminants. *Testing* is important for ensuring that drinking water meets standards for acceptable levels of contaminants *Water distribution* involves finding a way to safely provide households with an adequate and convenient supply of water. *Sanitation services* are critical for ensuring health, since otherwise it is impossible to keep water sources and households free from pathogens. Finally, some form of *sewage treatment* can protect downstream users and ecosystems from the harmful effects of human waste.

In the following sections, we explore the ways in which these tools have been used to provide healthy water.

## 9.2. Water and Sanitation: The Standard Model in Developed Countries

Most households in developed countries obtain their water from a piped supply delivered by a water utility. In the US, for example, only about

## Box 9.1. A Guide to Water-Related Diseases

Water-related infectious diseases are generally characterized by their mode of transmission, since this classification provides guidance on how to break the cycle of infection. The different categories and the most important diseases in each category are discussed below.

Waterborne ("fecal-oral"). These diseases are transmitted by contact with fecal material from infected people, most commonly through fecal contamination of drinking water. Additional transmission pathways include direct contact with human waste, transmission by flies, and contact with contaminated soil or household objects (fomites).

*Diarrheal diseases*: Diarrhea causes loss of fluids, nutrients, and electrolytes, and can lead to death from dehydration and low blood pressure. Diarrheal diseases include rotavirus infection, dysentery (bloody diarrhea often caused by *Shigella* bacteria or other bacteria, protozoa, or viruses), cholera (caused by the bacterium *Vibrio cholerae*), and typhoid fever (caused by bacteria of the Salmonella family). In developed countries, two protozoa, *Cryptosporidium parvum* and *Giardia lamblia*, both of which form cysts that are resistant to disinfection, are the most common causes of waterborne diarrheal diseases.

*Intestinal nematodes*: Several types of worms (the roundworm *Ascaris*, the whipworm *Trichuris*, and the hookworms *Ancylostoma* and *Necator*) can infect people, causing anemia, poor physical and mental development, and sometimes death. It is estimated that more than 2.5 billion people worldwide are infected, a decline from a high of ~3.5 billion people in the late 1990s. Transmission is primarily through contact with fecal-contaminated soil, where worm eggs can survive for up to 2 years.

Water-washed. A variety of infectious diseases, including trachoma, scabies, and leprosy, can be transmitted as a result of inadequate washing or by skin or eye contact with contaminated water.

*Trachoma*: Trachoma is a bacterial eye infection that can be transmitted through direct contact with secretions or indirect contact through household objects or flies. It can be prevented by adequate water and good hygiene practices. Some 6 million people currently suffer from the blindness that results from repeated or severe trachoma infections.

Water-based. Certain human parasites spend part of their life cycles living in water-based hosts; bathing in contaminated water can lead to infection.

*Schistosomiasis (bilharzia):* Schistosomes are intestinal worms that can cause chronic anemia, abdominal pain, and other symptoms. Infected people release large amounts of schistosome eggs in feces and urine. When that human waste ends up in a water body, the eggs hatch and infect snails living in the water. The schistosomes mature inside the snails and emerge as free-swimming microscopic life stages that can infect people who bathe in the water.

*Dracunculiasis (guinea worm):* In areas where dracunculiasis is endemic, the larvae of the guinea worm infect a copepod (the Cyclops water flea) in water sources. When the water containing these copepods is consumed, the larvae are released into people's bodies, where they mature to adult nematodes that cause painful blisters when they attempt to leave the body. Soaking of these blisters in water releases larvae into the water body, continuing the cycle of transmission. A concerted control effort has resulted in the near-eradication of the disease: the number of cases dropped from about 900,000 in 1989 to about 3000 in 2009, while the number of countries where the disease is endemic dropped from 20 to 6, all in Africa.

Water-related insect vectors. Several disease-carrying insects either breed in water (e.g., mosquitoes) or tend to bite near water (e.g., tsetse flies). Large water resource infrastructure projects, such as dams, can create stagnant water and increase the habitat for these insects.

*Malaria*: Malaria is caused by infection with *Plasmodium* protozoa and is characterized by fever, chills, nausea, and, in extreme cases, coma and death. The infection is transmitted by *Anopheles* mosquitoes, who obtain it by biting infected people. More than 300 million people become sick with malaria annually. The contribution of water management to the incidence of malaria is poorly understood, but dams (both large and small) and irrigation projects are thought to be significant factors. Prüss-Üstün et al. (2008) estimate that 42% of malaria incidence could be eliminated through environmental management. However, the primary approach for controlling malaria-carrying mosquitoes involves spraying of insecticides and/or use of insecticide-treated bed nets.

*Dengue*: Dengue is a viral infection characterized by fever, joint pain, and a rash. It is transmitted by *Aedes* mosquitoes, whose prevalence can be reduced by eliminating areas that collect shallow water, especially in urban areas.

*Japanese encephalitis*: This viral brain-swelling disease, restricted to Asia, is transmitted by *Culex* mosquitoes and can be reduced both by water management and by eliminating contact between mosquitoes and pigs, since the latter can serve to amplify levels of the virus.

*Onchocerciasis*: River blindness, mostly found in Africa, is caused by a nematode infection that is transmitted by black flies. Since the flies can breed in fast-moving natural waters, they are hard to control through water management; the main tool used to fight the disease is drug treatment.

*Lymphatic filariasis*: This nonfatal but debilitating nematode infection (closely related to the nematode that causes dracunculiasis) is transmitted by mosquitoes, but only when they breed in contaminated water, so it is more related to sanitation than to dams and irrigation projects.

Other

*Malnutrition*: Malnutrition can cause death and disability both directly, especially for children, and indirectly, by increasing susceptibility to infectious disease. Malnutrition is caused in part by water-related factors, such as repeated diarrhea or intestinal nematode infections. The complexity of sorting out the causes and consequences of malnutrition lends high uncertainty to attempts to estimate the fraction of malnutrition that is water related.

*Legionnaire's disease*: Legionellosis is transmitted by exposure to aerosol particles containing *Legionella* bacteria, which can grow in cooling towers and air-conditioning systems, such as found in hotels, cruise ships, and prisons. Symptoms include fever and coughing, and the disease can be fatal.

15% of the population obtains water from an individual private well. Likewise, most households in industrialized countries flush their waste into a sewer system that leads to some kind of centralized waste treatment. In the US, only about 25% of the population discharges to a septic tank instead. Thus, we refer to this centralized approach to water and sanitation as the "standard model."

## Water Supply

The water supply part of the standard model typically involves centralized water treatment and testing, as well as some efforts at source protection. Under the Safe Drinking Water Act of 1974 (SDWA), the EPA has established Maximum Contaminant Levels (MCLs) that must be met by public water supplies (Box 9.2). In addition, various rules promulgated under the SDWA specify the type of water treatment required, which differs depending on the size of the water utility, the type of source water (groundwater or surface water), and the levels of contaminants (especially cryptosporidium and other pathogens). Surface water generally needs to be treated by settling and filtration to remove solids, followed by disinfection (e.g., through addition of chlorine) to kill remaining pathogens, although communities can obtain a filtration waiver if they can show that source protection is adequate to protect water quality. Groundwater, which is more protected from contamination, usually only requires disinfection. Water utilities often add other chemicals as well, including fluoride for dental health (see section 9.5) and phosphate to prevent pipe corrosion.

The SDWA emphasis on filtration led to some perverse incentives to abandon source protection. In New Haven, Connecticut, for example, the water utility responded to the passage of the SDWA by attempting to sell off for development much of the land surrounding its reservoirs, in order to fund the newly required treatment plants. This would have led to increased pollutant inputs into the water sources (see section 6.2) and essentially traded off natural water protection for post hoc pollutant removal. Given the difficulties of removing pollutants, this would not have been a wise trade-off. Luckily, the State of Connecticut stepped in to prevent this scenario, by forbidding the sale of water supply lands and arranging for a semipublic entity to buy out the private water utility.

Several cities still rely entirely on source water protection, and have obtained waivers from the EPA to avoid building filtration plants. Most prominently, Boston and New York City have been successful, at least so far, in their battles to avoid spending billions of dollars on new filtration plants. New York's program to protect its water source in the Catskills has been held up as an example of effective large-scale watershed protection.

## Safety of Drinking Water

How effective has the standard model been at protecting human health?

Since 1854, when John Snow kicked off the "Sanitary Revolution" by demonstrating the linkage between contaminated water and a London cholera epidemic, waterborne diseases in the industrialized world have been dramatically reduced. However, they have not been eliminated. Between 1991 and 2002, there were 183 documented disease outbreaks in the US associated with drinking water (Reynolds et al. 2008), including the 1993 Milwaukee cryptosporidium outbreak, in which as many as 400,000 people were infected (MacKenzie et al. 1994). About 75% of the outbreaks (though not the Milwaukee one) involved groundwater, and

## Box 9.2. Setting Acceptable Levels of Contaminants in Drinking Water

In the US, the EPA is responsible, under the Safe Drinking Water Act, for establishing acceptable levels of chemical, radiological, and biological contaminants in drinking water. The EPA promulgates two types of standards: *maximum contaminant level goals* (MCLGs), which are set at levels that would protect against any adverse health effects; and *maximum contaminant levels* (MCLs), enforceable standards that may be higher than the MCLGs and that are determined by balancing health risks against the feasibility and cost of controlling contaminants.

MCLGs are generally derived from animal toxicity testing, with added margins of safety. For carcinogens (compounds capable of causing cancer), standard toxicological wisdom holds that there is no completely safe dose—any exposure causes some increased risk of cancer—so MCLGs are set at 0. For noncarcinogens, current models suggest a threshold exposure below which there is no toxicity, so the MCLG is set at that threshold (again, with a margin of safety).

Questions have been raised recently about whether this system is adequately protective of public health. Duhigg (2009d) calculated that MCLs for several contaminants—arsenic, chromium, tetrachloroethylene, and uranium—are high enough that they represent significant lifetime cancer risks—ranging from 1 in 600 to 1 in 16,666. It should be noted that these calculated cancer risks have a high degree of uncertainty, since they are derived by extrapolating downward from acute high-dose animal testing to chronic low-level exposure of people over the course of a lifetime; the nature of the dose-response curve at low levels is a matter of great controversy.

Ideally, our drinking water would be completely free from any potentially toxic chemicals. Yet there are real costs to developing and implementing MCLs. Distasteful though it may be, we must make choices as to where we should spend money to best protect public health. As Allan Freeze has pointed out, drinking-water MCLs have much greater cost per life saved than do some other public safety measures, such as clothing flammability and vehicle safety regulations (Freeze 2000). Again, the uncertainty as to what the risk really is may weaken that argument, but, paradoxically, reducing that uncertainty is prohibitively expensive in itself.

For biological contamination, there are many possible pathogens, so setting MCLs for each one is not generally feasible. Instead, the EPA approach involves setting MCLs for some selected pathogen groups that are particularly problematic (*Cryptosporidium, Giardia,* and *Legionella*), as well as some indicator groups that are not in themselves harmful but indicate a potential risk of fecal contamination (total coliform, heterotrophic plate count, turbidity).

None of the standards discussed above applies to private drinking-water wells, which in most states are not regulated at all. A recent USGS study of water quality in private wells (DeSimone 2009) found that 23% of the wells examined had at least one contaminant at concentrations higher than health guidelines (either MCLs or USGS-developed Health-Based Screening Levels for contaminants that don't have MCLs).

On the international scene, the WHO has published guidelines for drinking-water quality. While these are often applied wholesale to developing countries, they are, strictly speaking, just guidelines that countries may modify to suit their own particular conditions.

---

about 65% (though again not the Milwaukee one) involved individual wells or noncommunity water systems (systems that serve a seasonal or mobile population). In addition, developed countries also suffer from chronic (nonoutbreak) waterborne disease, though the amount is harder to estimate. Prüss et al. (2002) estimate that ~60% of intestinal illness in the US is water related, and Reynolds et al. (2008) estimate that 19.5 million illnesses occur each year in the US as a result of infection from drinking water, particularly among sensitive populations (young, elderly, immunocompromised). Further evidence of the importance of drinking water infection comes from a study in suburban Milwaukee (Redman et al. 2007), which found that pediatric emergency room visits for diarrhea increased after large rain events led to release of undertreated sewage into the drinking water source (Lake Michigan).

The risks associated with toxic chemicals in drinking water are the subject of some debate. On the one hand, most drinking water in the US meets the SDWA MCLs. On the other hand, these MCLs may not be adequately protective (Box 9.2). In addition, more than 20% of the nation's water utilities, mostly the smaller ones, have violated MCLs between 2004 and 2008, and more than 3 million Americans have been drinking water with illegal levels of arsenic or radioactivity (Duhigg 2009c). On top of this, the list of chemicals for which MCLs have been established is incomplete, and we are continually discovering new threats that need to be regulated. Recent examples include arsenic, for which a new MCL took effect in 2006; perchlorate, for which the EPA is still in the process of setting a standard; and dozens of other chemicals that have been found in drinking water sources but are not regulated. Lastly, our regulatory approach is not very effective at protecting sensitive subpopulations (such as children) or at dealing with the cumulative health risk posed by the mixtures of different chemicals that are potentially present at low levels in drinking water.

One of the trickiest issues in drinking water treatment is the problem of *disinfection by-products* (DBPs). Traditional disinfection with chlorine or bleach (hypochlorite) has been shown to lead to formation of small amounts of organochlorine by-products, which pose some cancer risk to people drinking the water. On the other hand, *not disinfecting* is clearly not a solution, since that would pose a much higher risk of waterborne disease. Instead, utilities try to find the right balance between the different risks. In particular, the goal is to apply enough disinfectant to eliminate even the most resistant organisms but still minimize formation of DBPs. This goal is easier to achieve if the source water is cleaner to start with, in terms of both pathogen levels and the organic matter that provides the starting material for DBP formation. The success or failure of Boston's and New York's source protection programs will probably ride on their ability to keep levels of indicator bacteria and organic matter in their source water low enough that they can achieve the dual goals of disinfection and minimal DBP formation.

Another approach to the DBP problem has been the use of alternative disinfectants. Several large utilities have switched over to the use of *chloramines*, which appear to cause lower levels of DBPs. However, this has led to unanticipated consequences. In particular, the absence of hypochlorite in the water can mean that metallic lead from water pipes begins to dissolve into the water. In a much-publicized public health problem, lead levels in Washington, DC, drinking water reached 10 times the acceptable levels as a result of the switch to chloramines, and the utility had to switch back to hypochlorite until it could remove the lead pipes.

Issues related to pipe corrosion continue to emerge. An incident of lead poisoning in Durham, North Carolina, in 2006 was traced to changes in the types of *coagulants* that were used in the treatment plant to remove organic matter (and thus reduce DBP formation). The new coagulant apparently led to pipe corrosion and very high levels of lead in tap water. An additional concern for water utilities is the skyrocketing price and limited availability of phosphate, one of the main chemicals added to drinking water to prevent pipe corrosion.

Pipes, treatment plants, and pumping stations in the US are aging and in need of maintenance and replacement. The American Society of Civil Engineers (ASCE) gives drinking water infrastructure a grade of D– (ASCE 2009), and EPA estimates a possible funding shortfall of $160 billion over the next 20 years (EPA 2002). One symptom of this problem is that approximately 15% of water introduced into the supply system is never delivered to customers (*unaccounted-for water*), much of which is lost to leaks (Hutson et al. 2004; ASCE 2009). Water infrastructure suffers from a serious visibility problem: since it is largely out of sight, people don't think about it as long as they can turn their taps on.

## Sanitation and Waste Treatment

The standard model for sanitation involves using water to flush human wastes from the home to a centralized treatment plant. The flushing itself eliminates the direct threat to the user, but only treatment eliminates the threat to ecosystems and to downstream users. Besides pathogens, domestic waste also carries high levels of biochemical oxygen demand (BOD), suspended sediment, nutrients, metals, and some organic pollutants. A conventional sewage treatment plant (STP) can include different levels of treatment:

*Primary treatment* involves screening and settling of larger particles.
*Secondary treatment* involves biological decomposition of much of the organic matter.
*Tertiary treatment* refers to any further treatment, most commonly for nutrient (N and P) removal.
*Disinfection* through chlorine, ultraviolet, or other treatments is designed to kill remaining pathogens.

The US Clean Water Act (CWA) mandated secondary treatment for all municipal sewage.

How effective has the standard model been at protecting ecosystem health? In the US, the widespread implementation of secondary treatment has resulted in huge improvements in the health of aquatic ecosystems. However, there are still some major shortcomings to this system:

- Large numbers of CSOs and SSOs still release untreated sewage into waterways, as discussed in section 8.6.
- There are still cities where STPs are only providing primary treatment (e.g., Honolulu).
- STP malfunctions and permit violations are common in some areas.

- Cumulative effects from multiple STPs discharging into one receiving water can lead to poor water quality, even if each STP is meeting its permit limits.
- Certain pollutants are not effectively removed by secondary treatment. Most prominent among these are nutrients, which commonly cause hypoxia and other water quality problems; and emerging organic pollutants, such as excreted pharmaceuticals.
- The amount of high-quality water used to flush away excreta is very large, which poses significant economic and environmental costs.
- Wastewater infrastructure—like drinking water infrastructure—suffers from deferred maintenance and system aging. It gets a D– from the ASCE (2009) and has an estimated funding gap as high as $150 billion over the next 20 years (EPA 2002).

## 9.3. Water and Sanitation in Developing Countries

### Millennium Development Goals

In September 2000, world leaders gathered for the Millennium Summit in New York, where they agreed to a set of eight goals for development activities, known as the Millennium Development Goals (MDGs). These goals generated 18 quantitative targets, including one that specifically deals with water and sanitation: "halve by 2015 the proportion of people without sustainable access to safe drinking water and basic sanitation."[1] The baseline date for this target is the year 1990, and the aim, it should be noted, is to halve the *proportion* of people without access. Of course, given population growth, even maintaining constant proportions of coverage means increasing the *number* of people covered.

Are we on track to achieve the water MDGs? Where do people in developing countries currently get drinking water? What kind of sanitation do they use? The Joint Monitoring Programme (JMP), jointly managed by the United Nations Children's Fund (UNICEF) and the World Health Organization (WHO), is responsible for answering these questions, which it does by compiling data from surveys that it carries out in different countries.

---

1. This was originally expressed as just a drinking water target; the sanitation component was added at the World Summit on Sustainable Development in Johannesburg in 2002.

The key terms of *access, safe drinking water,* and *basic sanitation* have been defined differently over the years. Most recently, the JMP has introduced drinking water and sanitation "ladders" describing the spectrum of practices that are used.

The drinking water ladder starts at the bottom with *unimproved* water sources that are highly susceptible to contamination, such as unprotected hand-dug wells, unprotected springs, and untreated surface water. Also included in this category are sources that are expensive or unreliable, such as bottled water and tanker trucks. The next rung of the ladder consists of *improved* sources that are outside the household and shared with others; this includes public taps and wells that are protected from contamination. The highest rung of the drinking water ladder is a *piped supply* into the house or yard. "Access to safe drinking water" is defined as including the top two rungs (i.e., improved sources, whether private or public).

The second-highest rung on the drinking water ladder is, in some ways, a problematic one. By the JMP definition, it includes high-quality water sources that are up to a kilometer from the home and provide as little as 20 liters per person per day. While having access to some safe water is better than not having access at all, the small quantities and the large distances involved do pose significant burdens, including the health risks of insufficient water for hygiene and of water recontamination during transport and storage.

The JMP's sanitation ladder begins with open defecation in fields or disposal of human waste with garbage. The second rung is unimproved sanitation facilities, basically latrines that do not effectively prevent contact between people and waste. These include *hanging latrines*, where the waste simply gets deposited on the ground or water surface below the latrine, and *bucket latrines*, where the user must empty the bucket of waste, often on the ground or into a water body. The third rung includes facilities that are hygienically acceptable but shared among two or more households, while the fourth and final rung consists of private improved facilities. Only the top rung is included in "access to basic sanitation," since shared facilities are likely to have problems with both maintenance and personal safety, especially at night.

The private improved facilities at the top rung of the sanitation ladder can be of several types. Besides flush toilets of the kind familiar to the reader, the facilities may be *pit latrines*, where the waste is deposited into a pit, which accumulates solid material and often lets liquid waste drain out into soil. When the pit is full, it must either be emptied and the material disposed of, or a new pit must be dug. Several improvements can be made to the basic pit latrine, including installation of ventilation and fly screening (*ventilated improved pit latrine*) and use of a small amount of water to flush waste through a short pipe to the pit (*pour/flush latrine*). The latter prevents odors by establishing a water seal and also allows waste to be directed to a new pit without relocating the latrine structure itself.

Note that the sanitation ladder deals only with sanitation, not with waste treatment. Even installation of flush toilets and a sewer collection system does not necessarily mean that the sewage receives treatment before being discharged to rivers or other water bodies.

Tables 9.1 and 9.2 show the latest (2008) data on drinking water and sanitation coverage for the world and by region. The drinking water picture is mixed (Table 9.1). The world as a whole is on track to meet the MDG, but the region that was in worst shape to start with—sub-Saharan Africa—is making little progress. Southern and southeastern Asia are making good progress but still lag behind other regions, especially in piped coverage. Overall, about 13% of the world (~880 million people) still lack access to safe drinking water, the vast majority of these (84%) in rural areas.

The sanitation picture (Table 9.2) is considerably gloomier than the drinking water picture, with about 38% of the world (2.5 billion people) currently lacking access to improved sanitation. While there are regions on track to meet the MDG, they are generally the regions that were in the best shape to start with. Sub-Saharan Africa, southern Asia, Oceania, and the world as a whole are not on track to meet the MDG. Southern Asia and sub-Saharan Africa have made some slow progress, but the task is daunting: two out of three people in those regions still lack improved sanitation, and one-quarter to one-half of the population still practices open defecation. Sanitation coverage is generally higher in urban areas, and about 70% of those lacking sanitation are rural dwellers.

## The Burden of Disease

What are the implications of the poor state of water and sanitation in developing countries? It is clear that there are strong links between water/sanitation access and health outcomes. Several interrelated factors are at work:

- Quantity: When water is expensive or must be carried long distances, the quantities used will be less than needed. This means that hygiene (washing hands, bathing, food hygiene) will suffer and the likelihood of waterborne and water-washed diseases will increase (see Box 9.1 for definitions).

TABLE 9.1

## Percentage of the population with access to different types of water supply

Data from JMP (2008). The final column indicates whether the region is on track to meet the 2015 target of halving (relative to 1990) the percentage of people without improved water supply. CIS = Commonwealth of Independent States. L Am = Latin America. SS Africa = sub-Saharan Africa.

| | | *Improved* | | *Total Improved* | | |
|---|---|---|---|---|---|---|
| | *Unimproved (%)* *1990 → 2006* | *Other (%)* *1990 → 2006* | *Piped (%)* *1990 → 2006* | *2006 (%)* | *2015 target (%)* | *On track?* |
| CIS | 7 → 6 | 22 → 21 | 71 → 73 | 94 | 97 | yes |
| L Am | 16 → 8 | 17 → 12 | 67 → 80 | 92 | 92 | yes |
| N Africa | 12 → 8 | 30 → 14 | 58 → 78 | 92 | 94 | yes |
| W Asia | 14 → 10 | 17 → 10 | 69 → 80 | 90 | 93 | yes |
| E Asia | 32 → 12 | 17 → 15 | 51 → 73 | 88 | 84 | yes |
| S Asia | 26 → 13 | 54 → 65 | 20 → 22 | 87 | 87 | yes |
| SE Asia | 27 → 14 | 57 → 54 | 15 → 32 | 86 | 87 | yes |
| SS Africa | 51 → 42 | 33 → 42 | 16 → 16 | 58 | 75 | no |
| Oceania | no data | no data | no data | 50 | 76 | no |
| WORLD | 23 → 13 | 29 → 33 | 48 → 54 | 87 | 89 | yes |

TABLE 9.2

## Percentage of the population with access to different types of sanitation

Data from JMP (2008). The final column indicates whether the region is on track to meet the 2015 target of halving (relative to 1990) the percentage of people without improved sanitation. CIS = Commonwealth of Independent States. L Am = Latin America. SS Africa = sub-Saharan Africa.

| | *Open Defecation (%)* *1990 → 2006* | *Unimproved (%)* *1990 → 2006* | *Improved Shared (%)* *1990 → 2006* | *Improved Private (%)* *1990 → 2006* | *2015 Target* | *On Track?* |
|---|---|---|---|---|---|---|
| CIS | no data | no data | no data | 90 → 89 | 95 | no |
| W Asia | 7 → 5 | 10 → 5 | 4 → 6 | 79 → 84 | 90 | yes |
| L Am | 17 → 7 | 10 → 8 | 5 → 6 | 68 → 79 | 84 | yes |
| N Africa | 16 → 4 | 17 → 14 | 5 → 6 | 62 → 76 | 81 | yes |
| SE Asia | 28 → 18 | 17 → 8 | 5 → 7 | 50 → 67 | 75 | yes |
| E Asia | 4 → 3 | 44 → 25 | 4 → 7 | 48 → 65 | 74 | yes |
| Oceania | no data | no data | no data | 52 → 52 | 76 | no |
| S Asia | 65 → 48 | 8 → 9 | 6 → 10 | 21 → 33 | 61 | no |
| SS Africa | 36 → 28 | 24 → 23 | 14 → 18 | 26 → 31 | 63 | no |
| WORLD | 24 → 18 | 17 → 12 | 5 → 8 | 54 → 62 | 77 | no |

- Quality: Drinking contaminated water increases the risk of waterborne disease, and bathing in contaminated water increases the risk of both waterborne and water-based diseases.
- Sanitation: Inadequate sanitation increases direct user exposure to fecal pathogens, as well as increasing the likelihood of indirect exposure through water, food, soil, or flies.

The most significant aspect of this water–disease linkage is diarrheal disease, which, according to WHO statistics, kills ~1.8 million people annually, or approximately the same as the number of people killed worldwide by HIV/AIDS. Practically all of these deaths are in developing countries (about 75% of them in Africa and Southeast Asia), and 90% of them are of children under the age of five. Diarrheal mortality has decreased slightly due to the use of oral rehydration therapy to prevent dehydration. However, at a more basic level, these diarrheal deaths are preventable by providing access to clean water, hygiene, and sanitation. Our collective inability to provide these basic necessities to so many is a massive human failure.

In addition to causing mortality, diarrheal diseases and other water-related diseases also cause significant suffering and disability. The standard measure used to quantify the overall burden of disease is the disability-adjusted life year, or DALY, which represents a lost year of healthy life. Data on water-related deaths and DALYs are presented in Table 9.3. For each water-related disease, estimates have been made of its total burden as well as the portion of this that is attributable to poor water practices. The latter represents the burden of disease that could be prevented by improving water, sanitation, and hygiene, or by better management of dams and irrigation projects (e.g., reducing standing waters to reduce habitat for malaria-carrying mosquitoes). The total impact of our poor water management represents about 8–9% of the total DALY burden from all causes. Even more dramatic, water-related mortality represents 25% of the deaths of children under 15. These are probably underestimates, since they include neither chemical contaminants nor diseases for which linkages to water are partial or not yet proven.

Have the improvements in water and sanitation access over the last two decades resulted in improvements in health outcomes? Unfortunately, the data to answer this question are not readily available, since WHO did not start publishing estimates of diarrheal mortality by country until the year 2000. However, the WHO data can be used to provide at least a rough answer to a related question: Do countries with better water and sanitation access have better health outcomes? Figure 9.1 shows current

TABLE 9.3
Estimates of annual water-related deaths and DALYs

Data for total deaths and DALYs are from WHO/UNICEF (2006); data for preventable deaths and DALYs are from Prüss-Üstün et al. (2008); both represent best estimates for the year 2002.

| | Deaths (thousands) | | DALYs (thousands) | |
|---|---|---|---|---|
| | Total | Preventable | Total | Preventable |
| **Waterborne** | | | | |
| diarrheal diseases | 1,800 | 1,500 | 62,000 | 52,000 |
| intestinal | | | | |
| nematodes | 12 | 12 | 2,900 | 2,900 |
| **Water-washed** | | | | |
| trachoma | 0 | 0 | 2,300 | 2,300 |
| **Water-based** | | | | |
| schistosomiasis | 15 | 15 | 1,700 | 1,700 |
| **Water-related insect vector** | | | | |
| malaria | 1,300 | 530 | 46,000 | 19,000 |
| dengue, Japanese encephalitis, onchocerciasis | 33 | 31 | 1,800 | 1,300 |
| lymphatic filariasis | 0 | 0 | 5,800 | 3,800 |
| **Other** | | | | |
| malnutrition | 260 | 860[a] | 17,000 | 36,000[a] |
| drowning | 380 | 280 | 11,000 | 7,900 |
| **Total water-related** | | | | |
| number | 3,800 | 3,200 | 150,000 | 130,000 |
| water-related as % of total from all causes | 6.7% | 5.7% | 10.1% | 8.5% |

[a] Preventable column includes estimates of deaths and DALYs from indirect, as well as direct, effects of malnutrition.

country-level data on diarrheal mortality and water/sanitation access. There appear to be fairly strong associations: countries with greater access are likely to have lower rates of diarrheal deaths, with water supply showing a slightly stronger relationship than sanitation.

However, there is a great deal of scatter in the relationships shown in Figure 9.1, indicating that many other factors (climate, access to health care, nutritional status, poverty, etc.) affect diarrheal deaths. Green et al. (2009) used a variety of such country-level variables to construct more sophisticated models of diarrheal disease burden. Their results suggest

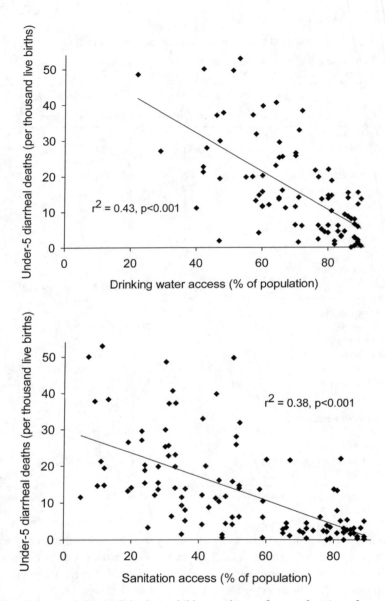

*Figure 9.1.* Annual diarrheal deaths in children under age five as a function of access to water (top) and sanitation (bottom). Each point represents a country. Countries with >90% access are not included. Data source: WHO Statistical Information System (www.who.int/whosis/en/index.html). Access data are mostly for 2006. Diarrheal deaths are calculated from total under-five deaths (2006 data) and % of under-five deaths caused by diarrhea (2000 data).

that water and sanitation access—especially rural sanitation access—are important in controlling the incidence of diarrhea. They predict that very substantial improvements in health outcomes will result if we are able to make progress in achieving rural sanitation access.

## Costs and Benefits

As noted above, we have made some progress in providing drinking water and sanitation access, but we still have far to go. Building and maintaining the infrastructure to meet the MDGs poses a huge logistical and financial challenge.

The WHO has recently provided updated estimates of the costs associated with this effort (Hutton and Bartram 2008). It estimates that extending new coverage to meet the MDGs will require expenditures over the next 10 years of somewhere between $2 billion and $14 billion per year for water and $6–$52 billion per year for sanitation, depending on what type of population is being served and what technologies are used. In addition, maintaining existing infrastructure in developing countries will cost somewhere around $54 billion a year.

This funding can potentially come from four sources: fees paid by those receiving the services; investments by national governments in their own infrastructure; international aid, either bilateral or multilateral; and for-profit investment, often by multinational water utilities. The first two options are clearly not adequate to provide the sums of money involved, especially for the poorest people in the poorest countries. Total international aid for water and sanitation averaged about $6.5 billion per year in the 1990s (JMP 2000), leaving a huge shortfall.

This funding gap has spawned two large movements. One calls for involving private for-profit companies in making these investments. We discuss this movement—and the backlash against it—in Chapter 12. The second outcome of the funding gap has been a search for lower-cost interventions that can achieve significant improvements in health outcomes, a phenomenon that we discuss in the next section.

It is important to point out that despite the high costs of providing water and sanitation services, the benefits of this investment are huge as well. Economic analyses have shown a positive benefit/cost ratio for meeting the MDGs and even for more expensive efforts such as ensuring *universal* access to safe drinking water and sanitation. These analyses estimate that for every dollar invested, we get returns of between $5 and $46 in benefits from reduced health care costs, increased productivity, and time savings (Hutton et al. 2007).

Despite this, there is a real financial problem:

- To the extent that the funding comes from international aid, the costs and the benefits accrue to different countries. Rich countries have their own priorities for their limited resources and don't necessarily give developing-country water and sanitation the attention that we might hope.
- To the extent that we expect developing countries to self-fund these projects, they often don't have the financial (or technical) resources to make these investments, even if they would provide great benefits in the long run; this is the vicious cycle of poverty.

## Reducing Water-Related Disease

As we try to move toward improved access and better health, we need to understand what types of interventions are most effective in breaking the linkages between water and disease. How should we invest our limited funding? Are there things we can do—short of the ideal of in-house access to unlimited high-quality water and sanitation—that can significantly improve health outcomes?

Most of the studies addressing these questions have dealt with diarrheal diseases. The possible transmission pathways for these diseases are complex and can involve (in the terms used in the classical "F diagram") food, fluids, fingers, fomites, fields, and flies. Interventions to break the cycle of transmission can be divided into four categories:

1. Water quality: water treatment is provided either centrally or at the household level, without providing access to a greater quantity of water.
2. Water supply: more convenient access to a larger volume of water is provided, without improvements to the quality of the water.
3. Sanitation: improved latrines are provided.
4. Hygiene: education and/or materials are provided to encourage hand-washing and other hygienic behaviors.

There has been particular interest recently in household-level water-quality treatment, referred to as point-of-use (POU) treatment. For example, the WHO has joined with a large number of nongovernmental organizations to form the International Network to Promote Household Water Treatment and Safe Storage. POU treatments includes a variety of techniques that individual households can use to improve the biological quality of their water, such as adding bleach; sunlight treatment; and specialized packets, often containing both a flocculant (to remove small particles) and a disinfectant. The excitement about this approach stems from three advantages:

1. Costs are quite low.
2. Since water is treated at the point of use, there is no opportunity for it to get recontaminated in transmittal and storage.
3. The approach is decentralized and gives individual households control over the quality of their water.

A number of individual studies have tested POU and other interventions using randomized controlled trials (RCTs), in which households were randomly assigned to either a treatment group (one or more of the interventions listed above) or a control group. The incidence of diarrhea was then noted (usually by self-reporting) in both the control and the treatment groups.

Fewtrell et al. (2005) conducted an influential meta-analysis in which they pooled the results from 38 individual RCTs. They noted a wide range of effectiveness among different studies, but found that, on average, all four types of interventions listed above performed about equally, reducing diarrheal incidence by 25–37%. Studies that included multiple interventions did not appear to be more effective than individual measures.

Fewtrell's study has influenced the thinking about water and health in several ways. First, the finding that individual interventions were as effective as multiple interventions lent support to the general idea that low-cost measures could provide good benefits even when funding is not available to provide full-blown access to water and sanitation. Second, the results suggested that simple hygiene measures could be surprisingly effective compared to more expensive infrastructural approaches. Third, the study appeared to support the excitement about POU treatments.

There has also been push-back against the POU bandwagon. Eisenberg et al. (2007) used a model to show that when multiple disease transmission pathways exist, interventions that deal with individual pathways (e.g., by focusing only on improving the quality of drinking water) are unlikely to be effective. As a result, they note that "when sanitation conditions are poor, water quality improvements may have minimal impact."

Schmidt and Cairncross (2009) conducted a meta-analysis of POU treatments and found an overall diarrheal risk reduction of 50%. However, they argue that this result may represent nothing more than *reporting bias,* since the vast majority of studies were not blinded. That is, the self-reporting of diarrheal symptoms is not an objective measure, and respondents may be influenced by the knowledge that they have received an intervention. The few studies that were blinded or used objective measures found no effect of the interventions. The same criticism could be applied to most of the studies in Fewtrell's meta-analysis.

Besides issues of bias, questions also remain about the sustainability and scalability of POU treatments. The studies mentioned above were generally short-term and did not look at the likelihood that households would continue using these treatments consistently after the aid groups left—including obtaining the necessary supplies. In addition, it is unclear how broadly acceptable these interventions are and whether they can be scaled up in a variety of countries. It should also be noted that POU treatments do not deal at all with the time burden of obtaining water and may in fact increase it.

Besides POU treatments (household scale), there is also great interest in decentralized interventions at the community scale. Especially in rural communities and urban slums, low-cost alternatives to the "standard model" may be quite effective, though they have not been studied systematically. The keys to success appear to be use of appropriate, low-cost, low-maintenance technology for water treatment and delivery; community ownership and management; immediate visible improvements in service and health; and education and training of local community leaders. International groups can jump-start this process, but its ultimate success depends on the local community.

## 9.4. Alternatives for Sanitation

When it comes to water supply, the standard developed-country model does pretty well, despite some of the issues discussed above, and I believe we would do well to eventually replicate its main feature—an in-house (or at least in-yard) supply of piped high-quality water—in the entire world. However, the situation is less clear when it comes to sanitation. The standard model does well at removing the health risk to the user, but it suffers from some major disadvantages:

- It requires a large investment in sewer infrastructure, which, because it is underground, is prone to failure and expensive to maintain.
- It uses a large amount of high-quality water for flushing away wastes.

- It leads to downstream pollution problems and high sewage treatment costs.
- It mixes together different waste streams, some of which are harmful and some of which are not.
- It treats as waste what is potentially a valuable resource—the organic material and nutrients in the excreta.

These disadvantages reflect a fundamental flaw in the standard model, which is that it involves *linear* one-way flows of water, nutrients, and energy.

The *ecological sanitation* ("ecosan") movement proposes a *circular* view of sanitation, in which water and nutrients are continually recycled (Langergraber and Muellegger 2005). This can be done in a variety of ways, but a basic principle of ecosan is separation of different waste streams so that each can be treated and reused in appropriate ways, usually at or near the site where it is generated:

- *Urine* is, in most cases, free of pathogens. It is also high in nutrients and typically contains ~90% of the nitrogen, ~50% of the phosphorus, and ~35% of the potassium in household wastewater. It can be used as an agricultural fertilizer with minimal treatment.
- *Feces* are high in pathogens and require high levels of treatment before reuse. However, household-level treatment is possible with techniques such as addition of ash to dehydrate the material and raise its pH, followed by high-temperature composting. It can then be used as a fertilizer, especially for nonfood products.
- *Household graywater*—water from showering, laundering, and so on—may contain low levels of pathogens, but can be reused without treatment for certain household uses, especially landscape irrigation and toilet flushing.

Although ecosan hasn't exactly replaced the standard model in developed countries, it has seen some limited implementation in Europe and the US. A recent *New York Times* opinion piece even extolled its virtue (George 2009).

One ecosan option is the No-Mix toilet, which separates urine from solid waste. Solids are flushed to the sewer in the usual manner, while urine can be stored and given to farmers for use as fertilizer.[2] A more

---

2. In some applications of the No-Mix toilet, urine is also delivered to the sewage treatment plant, but using less water and in a manner timed to balance out nutrient loads to the STP over the course of the day.

complete ecosan solution is offered by commercially available high-end composting toilets (www.biolet.com). These units, which generally require a power source, take up only slightly more space than a regular toilet and appear to be odor-free. The finished compost that is periodically removed can be used in landscaping.

Is ecosan a solution to sanitation problems in developing countries? In some ways, the ecosan approach is quite compatible with current sanitation practices in developing countries, which often involve collecting excreta on-site in pits. If ways can be found to safely treat and reuse this material, that would be a promising development. One option that is currently widely used in China and India for animal waste is anaerobic decomposition in biodigesters, with capture of the resulting biogas for use as a fuel. Addition of human waste to this system may be possible. Another alternative, in use in Nepal, involves a simple separating toilet, where both urine and feces are collected, treated, and applied on fields (Lamichhane 2007).

However, we need to be clear that ecosan should not compromise human health. As the pendulum swings back from the Sanitary Revolution of the nineteenth century—which saw human waste as a harmful substance to be flushed away—toward an appreciation of the possibilities of reuse, we should be careful not to let it swing too far. Techniques involving reuse, especially of fecal material, need to be tested in the field to ensure that all pathogens are inactivated, including the highly resilient eggs of intestinal worms like *Ascaris* spp.

In urban and periurban areas, high density may not allow safe on-site treatment, whether ecosan or other. In those cases, sewer systems need to be built to remove waste. However, the cost of building Western-style sewers is prohibitive in most developing countries, and several types of low-cost "simplified sewerage" are available. These involve smaller-diameter pipes that are installed at shallower depths and lower gradients.

If waste is collected with a sewer system, it should undergo some type of centralized treatment before release into the environment. Where building and operating a Western-style STP is too expensive, communities often use a series of *waste stabilization ponds*. Like STPs, these ponds involve both sedimentation and biological decomposition, but these processes occur much more slowly (over the course of weeks to months) and without added energy. Pathogen removal takes place through settling and oxidation of organic material and through sunlight inactivation.

## 9.5. Natural Contaminants

A variety of toxic chemicals pose potential threats to human health. As discussed in section 9.2, the standard model deals fairly well—though not perfectly—with this threat. In developing countries, however, poor regulation and monitoring of both chemical releases and drinking water quality mean that the threat of chemically contaminated drinking water is much higher. Communities downstream of large industrial or agricultural sources are especially at risk.

Natural contaminants can also pose a risk. Arsenic and fluoride are both naturally occurring toxic chemicals that present public health emergencies of the highest importance.

### Arsenic

Chronic exposure to moderate levels of arsenic can cause poisoning, known as arsenicosis, which can express itself in severe skin lesions, skin cancer, and other fatal illnesses. WHO guidelines recommend that drinking water contain no more than 10 μg/L of arsenic, although in high-arsenic areas (such as India), a standard of 50 μg/L is often used instead. Symptoms of arsenicosis typically do not appear until a person has been exposed for many years; once symptoms appear, providing uncontaminated water does little to alleviate them.

In 2006, the US drinking water standard for arsenic was lowered from 50 to 10 μg/L. The EPA estimates that the shift to the lower standard will lead to total annual benefits of $140–$198 million, including 21–30 avoided deaths from bladder and lung cancers. The costs of implementing the lower standard in the US are estimated to be $181 million per year (EPA 2001). Some would argue that even the 10 μg/L standard is not adequately protective: it is five times higher than the concentration that is estimated to represent a 1 in 10,000 lifetime cancer risk (EPA 2009b). Of course, there is a great deal of scientific uncertainty about these risk calculations, even though the arsenic risk assessment draws on human studies (particularly cases of high-level exposure in Taiwan) as well as animal models.

Arsenic occurs naturally at low levels in rocks, sediments, and groundwater throughout the world, but is found at much higher concentrations in certain locations. Two geological environments tend to lead to high arsenic levels. First, areas with alluvial (water-deposited) sediment, such as river deltas, can be high in arsenic, which, under certain conditions, can be released into groundwater through the chemical process of reduction.

In addition, elevated arsenic levels can be found in arid inland regions with high rates of weathering and evaporation. Countries with areas of documented high-arsenic groundwater include the US, Canada, Mexico, Argentina, Hungary, China, and Vietnam.

The center of the arsenic disaster is the delta of the Ganges-Brahma-putra-Meghna (GBM) rivers in Bangladesh and the nearby Indian district of West Bengal. It is in this region that doctors in the early 1980s started noting large numbers of patients with skin lesions and were able to trace the problem to high arsenic levels in groundwater. Researchers slowly began to understand the magnitude of the problem and attempted to bring it to the attention of national authorities and the international public health community. However, these warnings were largely ignored, and in 1992, when the British Geological Survey (BGS) tested water quality in wells in the area, they neglected to include arsenic in their analyses. The BGS was later sued (unsuccessfully) for negligence.

A conference in 1995 started to bring more serious attention to the issue, and there have been increasing attempts (discussed below) to provide safer drinking water to the area. However, this "largest mass poisoning in history" (Smith et al. 2000) continues to unfold, with estimates that, as recently as the late 1990s, approximately 30 million people were drinking water with arsenic concentrations above 50 µg/L (Chakraborti et al. 2002). Because of the long latency period, the number of people who will eventually suffer or die from arsenicosis is unknown.

Ironically, the arsenic problem in Bangladesh can be traced to the efforts by UNICEF in the 1970s to provide alternatives to the biologically contaminated surface water that had been serving as the main source of drinking water. UNICEF installed large numbers of tubewells—narrow-diameter wells equipped with a pump—that tapped the arsenic-contaminated aquifer. (Due to the complex geology of the region, highest arsenic levels are found at intermediate depths (~30–200 m), with both shallower and deeper aquifers being relatively low in arsenic.) The convenience and health benefits of the UNICEF wells inspired many people to install private wells, and the total number of shallow tubewells in Bangladesh is now more than 12 million (Hug et al. 2008). Since 1997, efforts have been made to test these wells for arsenic and paint them green (safe) or red (unsafe), though some of the test kits used for this purpose are not very accurate. Certain districts are much more likely to have contaminated wells, but even within a village, different wells can have very different levels of arsenic, due to the patchy occurrence of the geological conditions that favor arsenic release.

Several strategies are being implemented in Bangladesh to reduce the population's exposure to arsenic: encouraging people to switch to wells that have been painted green; installing deeper tubewells or shallower dug wells that tap lower-arsenic waters (though the shallower wells may be more susceptible to surface sources of fecal contamination); and providing various types of household-level treatment to remove arsenic from water. The challenge of devising and distributing a simple, effective, low-cost arsenic removal kit has not yet been met, though there are several promising technologies. In Vietnam, where another 12 million people are drinking high-arsenic water, these kits have been more successful due to differences in the groundwater chemistry (Hug et al. 2008).

Recently, concerns have also been raised globally over human exposure to arsenic from rice consumption, especially for populations that consume large quantities of rice or consume rice that is particularly high in arsenic. Rice appears to be particularly susceptible to arsenic accumulation due to its biology. In addition, much of the world rice crop is grown in high-arsenic areas, including delta sediments in Bangladesh, as well as former cotton fields in the US where arsenical pesticides had previously been used.

## Fluoride

Fluoride provides a good illustration of Paracelsus' maxim that "the dose makes the poison." At low doses, fluoride appears to protect against dental caries (cavities), and many water suppliers add fluoride to drinking water in order to provide this benefit (although the evidence to support this health benefit is relatively weak, and fluoridation is still controversial in many places). At higher doses, fluoride causes *dental fluorosis*—characterized by mottling and structural damage to the teeth—and *skeletal fluorosis*—characterized by increased bone density, with associated immobility, pain, and in extreme cases, deformities and paralysis.

The line between the beneficial and harmful effects of fluoride appears to be remarkably thin. Optimal levels in drinking water are said to be in the range of 0.7–1.2 ppm, while levels above 1.5–2.0 ppm can cause fluorosis (Ozsvath 2009).

Surface water is generally low in fluoride, but the concentration of fluoride in groundwater can vary tremendously, depending on geology and climate. In most countries, little detailed information exists on the extent and location of high-fluoride drinking water and the number of people exposed. Ayoob and Gupta (2006) estimate that perhaps 200 million people in at least 25 countries suffer from fluorosis problems due to naturally high concentrations in groundwater, while Fawell et al. (2006) calculate that China and India each have about 25 million people with water-related fluorosis.

Fluoride is harder to remove from water than arsenic, although there are some low-cost technologies that have had limited success. Intervention to reduce the incidence of fluorosis more typically involves efforts to provide alternative sources, such as introducing rainwater harvesting techniques or identifying groundwater that is lower in fluoride. Providing better nutrition is also part of the picture, since the effects of fluorosis are exacerbated by poor nutritional status.

## 9.6. Bottled Water

In both developed and developing countries, bottled water can represent an alternative source of drinking water. Bottled water sales have grown remarkably rising from about 73 MCM $yr^{-1}$ in 1996 to 169 MCM $yr^{-1}$ in 2004 (Gleick 2006). The 2004 total represents an average of 24 liters of bottled water for every person on the planet. The countries with the highest rates of per capita consumption are Italy, Mexico, and the UAE (all more than 160 L $person^{-1}$ $yr^{-1}$), with the US coming in 11th at 91 L $person^{-1}$ $yr^{-1}$ (Gleick 2006).

The rise of bottled water is a deeply disturbing phenomenon. Compared to a piped water system, bottles are a tremendously inefficient way to deliver water—especially when the piped system already exists! In addition, bottled water diverts resources that could be used to improve and extend the piped system. And, of course, it results in the production and disposal of huge volumes of plastic.

Why do people buy bottled water, when it is so much more expensive than tap water (usually by a factor of at least 1000)? Gleick (2006) presents three explanations: water quality, convenience, and taste.

For households in developed countries, the water quality argument is generally a mistaken one: the quality of the water coming out of the tap is usually at least as good as that purchased in a bottle. In fact, water quality standards and testing requirements are generally stricter for tap water than for bottled water, and there have been numerous cases of biological and chemical contaminants in bottled water. In addition, there is good evidence that water bottles leach an endocrine-disrupting chemical known as bisphenol A (BPA). Of course, as noted above, not all tap water is safe, even in industrialized countries, and there are certainly cases where private wells or municipal supplies are contaminated; in such cases, bottled water may be the best choice until improvements can be made to the piped supply.

For households in developing countries without access to a safe supply of pathogen-free water, bottled water may represent the only reasonable source of water—though it still may not be as safe as it should be. In any case, it clearly makes sense for society to invest in providing household-level access rather than bottled water. The $100 billion or so that is spent every year on bottled water could go a long way toward meeting the needs discussed in section 9.3.

The convenience factor is, no doubt, a large part of the bottled water picture, especially given the high level of daily mobility in many countries. This will always be a factor, but change appears to be coming. More and more people can be seen carrying their own reusable water bottles, and environmental groups have successfully made plastic water bottles a mark of shame.

Taste can sometimes be a real consideration, especially when tap water is heavily chlorinated. However, taste tests often reveal that people can't actually tell the difference between bottled and tap water.[3]

## 9.7. Household Water Conservation

In industrialized countries, where households generally enjoy a cheap supply of piped water, excessive and wasteful water use can contribute to water scarcity and degradation of aquatic ecosystems. In poor countries as well, precious water is often lost to leaks and inefficient end uses. Are there ways to reduce household use, especially when that use appears to be excessive?

### Current Use

How do we define excessive use? Or, put another way, what is a reasonable target for household water use? One way to start answering this question is by looking at the range in current residential water use in industrialized countries (Table 9.4). The vast differences in water use per person are, to some extent, unavoidable consequences of climate, geography, culture, and infrastructure. Yet countries with the highest water use, like the US, can undoubtedly do a great deal to bring their use down closer to the norm.

We can explore geographic variation at a finer scale by looking at residential water use in different cities in the US (Table 9.5). People in different parts of the country tend to use very different amounts of water,

---

3. Adding lemon to tap water can both mask the chlorine taste and remove the chlorine through the chemical reaction of the chlorine (an *oxidant*) with the ascorbic acid (vitamin C, a *reductant* or *antioxidant*) found in the lemon juice.

TABLE 9.4
Country-level domestic water use for industrialized countries

Data source: AQUASTAT (2009). Data shown are for the top 5 and bottom 5 water users among the 31 countries with per capita GDPs >$25,000. These country-average numbers include not just residential water use but also commercial and institutional use, as well as leaks. They also suffer from the quality problems associated with globally reported data. The number for the UK is ~40% lower than the residential water use reported for the UK by Vickers (2001).

| Country | Average Domestic Water Use (Lpcd) |
| --- | --- |
| minimum water requirement | 50 |
| Netherlands | 83 |
| UK | 95 |
| Singapore | 101 |
| Finland | 179 |
| Ireland | 181 |
| industrialized country median | 317 |
| US | 573 |
| Qatar | 580 |
| Bahrain | 660 |
| New Zealand | 708 |
| Canada | 787 |

with cities in the Northeast and Pacific Northwest having the lowest water use, followed by the Southeast and finally the West. This result may seem surprising given the greater water scarcity in the West, but it is driven primarily by differences in outdoor water use, with households in the arid West using large amounts of water for landscaping, due to the long growing season and the low rainfall.[4] Other factors also contribute to regional and household-to-household differences in water use:

- Wealth/house size: More affluent households tend to use more water, both outdoors (where they have larger lawns as well as pools or other water features) and indoors (where they have more and larger plumbing fixtures).
- Building type: Multifamily buildings tend to consume less water per person, due to shared plumbing and outdoor space.
- Building age: Older buildings have leakier pipes and older, less efficient fixtures.
- Price: As we discuss in Chapter 12, lower prices encourage greater water use. The Northeast (with its low water use) has the highest average water and wastewater charges in the US.

---

4. An additional factor in some parts of the West is the household use of evaporative cooling devices, which can use large amounts of water.

TABLE 9.5
Residential water use for selected US cities

Cooley et al. (2007) and Cooley and Gleick (2009) data are for single-family residential use only. Vickers (2001) data are for single-family and multifamily residential use. All data represent the late 1990s/early 2000s.

| City | Average Residential Water Use (Lpcd) | Data Source |
|---|---|---|
| Manchester, VT | 210 | Vickers (2001) |
| Seattle | 240 | Cooley and Gleick (2009) |
| | | |
| Boston | 260 | Vickers (2001) |
| Atlanta | 340 | Cooley and Gleick (2009) |
| | | |
| Tampa | 370 | Vickers (2001) |
| Albuquerque | 420 | Cooley et al. (2007) |
| Tucson | 430 | Cooley et al. (2007) |
| Los Angeles/San Diego | 470 | Cooley et al. (2007) |
| Las Vegas | 620 | Cooley and Gleick (2009) |
| | | |
| Denver | 640 | Vickers (2001) |
| Phoenix | 830 | Vickers (2001) |
| Las Virgenes, CA | 950 | Vickers (2001) |

Given the limitations of geography, climate, and other factors, it is not realistic to expect the entire country to reduce its water use to the level of Manchester, Vermont. But the variations within a region make it clear that there is untapped conservation potential in cities in every part of the country.

Another way to understand the potential for reducing household water use is to take a closer look at how water is currently used in a typical home in the US. As noted above, outdoor water use can vary tremendously but on average makes up about 30% of residential water use. The distribution of indoor water use, based on a study of 11 North American cities (Mayer et al. 1999), is shown in Table 9.6. These data illustrate the tremendous amount of water that goes to flush toilets, run clothes washers, and feed leaks: these three categories make up 62% of water use in a typical nonconserving home. At the same time, the data also illustrate the great potential to reduce these water uses (by about half!) by moving to more efficient fixtures.

TABLE 9.6
Average indoor water use in standard and conserving
single-family homes (Lpcd)

Data for standard homes are from Mayer et al.'s (1999) study of 11 North American cities. Data for
conserving homes are estimated by Vickers (2001).

|  | Standard Home | Conserving Home |
|---|---|---|
| toilets | 70 | 31 |
| clothes washer | 57 | 38 |
| showers | 44 | 33 |
| faucets | 41 | 41 |
| leaks | 36 | 15 |
| baths | 5 | 5 |
| dishwasher | 4 | 3 |
| other | 6 | 6 |
| total | 263 | 172 |

## Tools for Conservation

How can a household or a city reduce its water use? Here are some of
the most important tools available for conservation, or "demand management," particularly in industrialized countries.

Efficient fixtures: Advances in technology and design have allowed the
creation of plumbing fixtures and appliances that do the same job with
less water. The Energy Policy Act of 1992 introduced maximum water-
use requirements for toilets, urinals, showerheads, and faucets installed
after 1994, but older fixtures are still in place in some households. The
replacement of older 3.5 gpf (gallon per flush) toilets with 1.6 gpf toilets
(the maximum allowed under the act) can save considerable water. Some
newer toilets use still less water: the EPA gives the WaterSense label to
toilets that use 20% less than the 1.6 gpf standard, and dual-flush toilets
and waterless urinals are also becoming more common. There are no
national water-use requirements for clothes washers or dishwashers, but
water-efficient units are available and are encouraged by EPA's voluntary
WaterSense and Energy Star programs and by new energy use standards.
In calculating payback periods for replacing older fixtures with more
efficient ones, it is important to include not just the cost of the water
saved but the cost of the energy used to heat that water, as well as other
ancillary benefits.[5] Water utilities can encourage fixture replacement by

---

5. For example, front-loading clothes washers have benefits in at least four areas:
lower water use, lower energy use, less wear on clothes, and a faster spin cycle that

providing rebates or distributing smaller items (e.g., low-flow shower-heads) for free.

Behavioral change: Many people, even in water-scarce regions, are not aware of how much water they are using. Increasing awareness of water scarcity can lead to short- or long-term behavioral change, such as shorter showers, fewer loads of laundry, less lawn-watering, and so on. On the other hand, behavioral change can work in the opposite direction as well. In particular, efficient fixtures don't always reduce water use by the expected amount, since people may respond to the greater efficiency by increasing their use (the *rebound effect*). For example, Davis (2008) found that households that were given efficient clothes washers increased their clothes washing by 5.6%.

Pricing: Increasing the price of water can motivate people to reduce their water use by both technological and behavioral means. In the US, water is generally the cheapest utility that people pay for, and consequently the most ignored. The pros and cons of higher water prices are discussed in detail in Chapter 12.

Outdoor water use: Given the large volumes and nonessential nature of outdoor water use, targeting this category makes good sense. Landscape watering is particularly a problem in the arid West, as noted above, and in suburban households with large, manicured lawns. Replacing lawns with drought-tolerant native species can eliminate most outdoor water use. Where lawns are considered necessary, they should be allowed to grow relatively tall and should be watered sparsely. Automatic irrigation systems should be equipped with rain sensors, to avoid the use of irrigation when it is not needed.

Leaks: Finding and fixing leaks, both at the household level and at the utility level, can save a great deal of water. For older water systems, *unaccounted-for water* can amount to more than 20% of water use; in addition to leaks, this includes water that is being metered incorrectly, water used by the utility for system maintenance, and sometimes firefighting water. The American Water Works Association (AWWA) recommends that unaccounted-for water should be <10% of water produced.

---

produces a load that is not as wet and thus requires less time in the dryer.

Graywater: Water draining from sinks, dishwashers, clothes washers, and showers is not highly polluted and can potentially be used again within the household rather than sent to an STP for treatment. The main uses for this graywater tend to be either toilet flushing or landscape irrigation. Local health codes sometimes make it difficult to install a graywater system, but many thousands of these systems are in use without any health problems. Opinions on the utility of this approach vary, with the water conservation specialist Amy Vickers stating that "a graywater system offers few practical benefits for most single-family homeowners" because of its high cost and significant maintenance needs, as well as the small volumes of water involved, especially after installation of efficient fixtures (Vickers 2001).

Waterless toilets: Composting toilets and urine-separating toilets can minimize or eliminate the use of water for flushing away wastes. Both graywater use and waterless toilets can be part of the "ecosan" approach discussed in section 9.4.

Rainwater harvesting: As discussed in Chapter, 7, households can capture the rain that falls on their house and property and store it for future use. Most commonly, this means using a rain barrel to capture roof runoff and store it for watering gardens, but it can also include larger-scale systems that store rainwater for indoor use. In some western states, rainwater harvesting is technically illegal, since the rights to the rainwater are owned by whoever owns the nearby surface water or groundwater rights.

Limits on growth: In areas that are experiencing rapid population growth and suburban sprawl, all the conservation measures discussed above may not be enough to counteract the upward trend in water demand. As more and more people move into an area and build large houses surrounded by huge expanses of turfgrass, they may stress the water supply to its limit. There are currently very few mechanisms to align land development with water availability. This needs to change, and is starting to.

## Effectiveness

What does this all add up to? Can conservation programs really reduce water use enough to make a difference? The data in Table 9.6 show that water savings can amount to about a third of indoor water use when conservation measures are applied at the household level. But how does this play out at the more relevant scale of an entire city or utility?

The experience of several cities that have implemented concerted conservation programs suggests that savings from conservation can be quite significant and can potentially alleviate the need for additional water supply.

Austin, Texas, started experiencing water supply problems in the early 1980s, when rapid growth and high per capita water use began to strain infrastructure capacity. Also looming on the horizon was the threat of running out of water, since Austin has fixed water rights from the Colorado River of 65 MCM yr$^{-1}$, which it was projected to exceed sometime in the early decades of the twenty-first century. As a result, Austin has implemented a series of conservation programs over the last 25 years that have resulted in a decrease in per capita water use from 770 Lpcd in 1980 to 620 Lpcd in 2003 (still well above the national average). Here are some key components of these programs:

- Toilet replacement programs, which offered free toilets and even covered installation costs (though there were maintenance difficulties with some of the toilet models installed)
- Strict guidelines for irrigation systems, with technical assistance to help homeowners comply and fines if they don't
- Xeriscaping programs to encourage replacement of turfgrasses with native plantings that require much less water

Las Vegas, Nevada, is well known for being a profligate water user in the midst of a desert. However, as its rapid growth has strained its water supply, city planners have included demand management in their response (along with increasingly far-reaching plans to augment supply). The most widely celebrated aspect of Las Vegas' program is the turfgrass buyback program, in which homeowners are paid $1.50 per square foot of turf destroyed. Cooley et al. (2007) estimate that more aggressive conservation measures could decrease water demand by 30–40% and reduce the need to seek new water sources.

Seattle, Washington, has managed to decrease its total water consumption even as its population has grown, with an impressive decline in per capita consumption from 600 Lpcd in 1975 to 350 Lpcd in 2008.[6] Seattle's comprehensive water conservation program targeted all opportunities for reducing water use, including replacing fixtures, modifying landscaping practices, raising prices, and improving system operations.

---

6. Note that this number is higher than that shown in Table 9.5 because it includes all water delivered by the utility, not just that used by residential customers.

Lastly, Gleick et al. (2003) conducted an analysis of the potential for urban water conservation in the entire state of California. They estimated that urban water use could be reduced by 34% using existing technologies or by 29% using only technologies that are cost-effective, that is, cheaper than the costs of new supply ($0.49/m³). These water savings would come from 25–40% reductions in all sectors: commercial/institutional and industrial as well as indoor and outdoor residential.

## 9.8. Conclusion

In this chapter, we discussed household water and sanitation issues in both rich and poor countries. Most (but not all) people in the US and other rich countries benefit from a cheap, safe supply of household water and a cheap, safe sanitation system, which bring enormous, underappreciated dividends to our health and standard of living. At the same time, there are certainly issues to work on: minimizing and mitigating instances of drinking water contamination, balancing risks and costs to better protect public health, better maintaining vital water and sanitation infrastructure, improving sewage treatment, and making better use of each drop of water in order to alleviate scarcity and environmental degradation.

With all the concern over the safety of drinking water in the US (e.g., Duhigg 2009c, 2009d), we need to keep these issues in perspective. Saving 30 lives annually in the US (the estimated benefit of the new arsenic standard) is a good thing, but we must also bear in mind the 1.5 million children who die each year in developing countries from preventable diarrheal diseases. Meeting the Millennium Development Goals of providing water and sanitation access is a large task, but one that will bring enormous benefits. Creative thinking and action toward achieving these goals is an urgent imperative.

We turn next to another vital water-using activity—agriculture.

# 10

## Crops and Drops: Getting More from Less in Agricultural Water Use

In this chapter, we turn to the agricultural sector, with an eye both to understanding its role in the global water crisis and to thinking about ways to move toward a more sustainable future. We begin the chapter by setting the context: agriculture requires a great deal of water, and several important practical and political consequences stem from that fact. We then summarize the recent history and current state of agricultural development, discuss ways to increase the productivity of agricultural water, and finally outline the issue of agricultural pollution.

### 10.1. Agriculture as a Water User

Two basic facts are critical for understanding agricultural water use: (1) the agricultural sector is a huge consumer of water, and (2) water used in agriculture generally has a relatively low economic value.

As discussed in Chapter 3, agriculture—specifically irrigated agriculture—is by far the biggest user of blue water, both globally and in most countries. If we include green water as well, agriculture—now including rain-fed crops and pastureland—becomes an even more dominant force in the human appropriation of the global water cycle.

At the same time, the water used by agriculture usually has a lower value than the same amount of water used for domestic or industrial purposes. In the case of household use, a minimal amount of water to sustain health has an extremely high value, and less compelling household uses, such as lawn watering, are often highly valued as well. Similarly, industrial users often are willing to pay high amounts for the water that they need to produce their high-value products. In contrast, farmers—at least in the current economic climate—produce a product that is relatively cheap, and thus they can't afford to pay very much for water. Two specific examples illustrate the pattern (Brewer et al. 2008): In California, an acre-foot of water used to grow cotton and alfalfa generates $60 in revenue, while an acre-foot used by the semiconductor industry generates $980,000. And Imperial Irrigation District (IID) farmers pay about $15 per acre-foot of water, while San Diego was willing to pay $225 per acre-foot for that same water. Of course, food is basic to life (almost as basic as water is!), and if we were ever to face severe global food scarcity, the price of agricultural products—and the value of water for agriculture—would go way up. But that is not currently the case.

From these two facts—the high volume and low value of agricultural water—stem three important corollaries:

*Other users tend to turn to agriculture as a potential source of water.* As cities grow (in both population and standard of living) and need more water— and as local natural water sources are fully tapped—urban water managers tend to cast an eye on the vast volumes of water used by farms, often a relatively short distance from expanding conurbations. Transfers of water from farms to cities are sometimes voluntary and transparent: the value of the water to urban users is high enough that farmers are willing to produce and sell water, instead of food. In other cases, these "water grabs" are filled with subterfuge and conflict, as in the infamous diversion of water to Los Angeles from the farming community of Owens Valley in the early 1900s.

Similarly, as environmental groups look for sources of water to restore rivers and other aquatic ecosystems, they most often turn to agricultural allocations, complaining of inefficiency, perverse subsidies, and overuse in the agricultural sector.

An example of the way that agricultural water is viewed in a situation of scarcity is provided by the response of Israeli water managers to drought. Israel has a highly centralized water allocation system, and allocations to farmers are the first to be cut in a drought, while, until recently, urban users were merely encouraged—with greater or lesser success—to adopt water-saving measures.[1] As Fischhendler (2008a) points out, how-

---

1. A recent large increase in household water prices represents somewhat of a change in this dynamic.

ever, agriculture has its own political power, and allocations to Israeli farmers during drought years often end up being much more than they should be (and more than originally announced), with the difference being made up by depleting storage.

*Relatively small changes in agricultural water use efficiency can have large implications for global and local water budgets.* Given the great disparities in water use efficiency within the agricultural sector, and given the large amounts of water involved, there is the potential to free up a great deal of water by moving toward more efficient technologies. At the global scale, the *range* among different scenarios in projected agricultural water withdrawals for the year 2050 is approximately 1200 km³ yr⁻¹ (CAWMA 2007)—more than all current nonagricultural withdrawals combined. In other words, which path we choose for the future of growing food will make a huge difference in the amount of water available for other uses, whether those are household, industrial, or environmental. Much of this chapter will be devoted to discussing ways that we can move toward an agriculture that is less water intensive.

*Trade in virtual water can potentially alleviate both local and global water scarcity.* If you take food out of the equation—by importing it, rather than growing it locally—even the most water-scarce countries have enough water to meet their population's basic needs for household and industrial uses. As discussed in Chapter 7, food trade plays a significant role in alleviating water scarcity for the most arid countries, as well as in saving water globally by growing food in locations that require less water. The phenomenon of virtual water trade is a direct consequence of the fact that growing food requires large volumes of low-value water: agriculture tends to shift away from regions where water is scarce and where there are other, more high-value demands for that water.

Yet there are certainly many counterexamples, where virtual water seems to flow backward: arid regions where we are using huge volumes of water to grow water-thirsty crops, often for export to more humid regions and at the expense of higher valued uses. These misguided activities tend to be driven by an interlocking combination of factors: artificially low water prices; a need for foreign currency, historical water allocations that are locked in and hard to change, inadequate recognition of the value of water for ecosystems, and cheap arable land with ideal growing conditions (other than the absence of water!).

## 10.2. The Green Revolution and Its Limits

The second half of the twentieth century saw extraordinary changes in our ability to grow food, a phenomenon known as the Green Revolution. As world population doubled from 3 billion to 6 billion over the period 1961–2000, food production more than kept pace.[2] World cereal production, for example, increased by a factor of 2.3 over this time period. Most remarkably, this increased production took place without significant increases in the total amount of land devoted to agriculture, but was instead a result of higher yields per unit of land (Figure 10.1). Three primary factors contributed to the success of the Green Revolution:

1. The introduction of new, high-yielding varieties of staple crops (HYVs) that have a higher harvest index (the fraction of the biomass that is edible) and better overall growth
2. The use of chemical inputs: fertilizers to increase plant productivity and pesticides to control weeds and insects
3. The expansion of irrigation

Yet the world food picture is not entirely rosy:

- The higher yields of the Green Revolution have not yet reached the entire world; most of Africa continues to experience low yields, on the order of 1 ton per hectare for cereal crops (Figure 10.1).
- There are signs that yields may be leveling off in the areas that have seen the greatest growth in yields. There is particular concern that rice yields in Asia may be flat-lining, and projections for the next 50 years suggest that yields of most cereal crops will show little growth (from their already high levels) in developed countries. There is also some evidence that climate change (especially higher nighttime temperatures) may be having a negative effect on rice yields (Peng et al. 2004).
- Hunger and malnutrition are still large problems, with around 800 million people globally suffering from undernourishment. This is not caused by global shortages of food but by the fact that many impoverished people can't afford to buy adequate food or obtain the technology that would improve their ability to grow their own food.

---

2. The Green Revolution began around 1945; I cite the period beginning in 1961 because data on global food production are not available before that time.

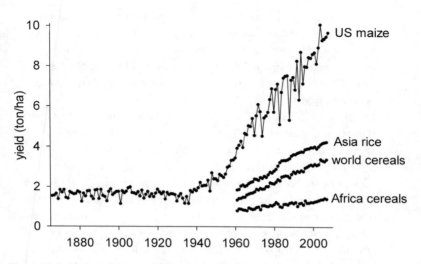

*Figure 10.1.* Changes in crop yields over time. US maize data 1866–2008 from USDA (www.nass.usda.gov/), other data 1961–2007 from FAOSTAT (faostat.fao.org/).

- Land and water are becoming limiting factors. In order to feed growing populations, we will need more arable land and blue or green water. Where will this land and water come from?

In addition, each of the three pillars of the Green Revolution is now showing its limitations:

1. HYVs: HYVs have been criticized for requiring intensive water and chemical inputs in order to reach their potential. In addition, further genetic manipulation may not continue to increase yields, since there is a natural limit to plant productivity. At the same time, HYVs have not really taken off in most of Africa, in part because of the cost of seed and in part because these varieties were not designed for African growing conditions. A new generation of HYVs—exemplified by New Rice for Africa (NERICA)—has the potential to bring the Green Revolution to the world's poorest continent.

2. Chemical inputs: While the Haber-Bosch method for manufacture of nitrogen fertilizer may have been one of the most life-saving scientific discoveries of the twentieth century, the overuse of chemical fertilizers has also had serious ecosystem impacts. In particular, runoff of nitrogen and phosphorus from agricultural lands has been responsible for the overfertilization of many

inland and coastal aquatic systems, resulting in a growing number of hypoxic "dead zones" and other harmful impacts. Likewise, pesticide use has had toxic effects on humans, birds, and
other nontarget species. As was the case for HYVs, the benefits
of fertilizer inputs have not yet reached much of Africa, largely
because of cost.

3. Irrigation: Much of the increased food production of the Green
Revolution came from land that was equipped with irrigation
as its primary water source, especially in Asia, which has ~70%
of the world's irrigated land. Irrigation water can come from a
variety of sources, but the two most common are groundwater
pumped by individual farmers and large surface water projects
that deliver water from a large dam to a massive command area
using hundreds of miles of canals. Each of these has social and
environmental impacts, as discussed in sections 5.3 and 7.3, respectively. The amount of irrigated land doubled over the period 1961–2000 but recently shows signs of leveling off, as new
sources of irrigation water become harder to develop.

Irrigated agriculture generally has higher productivity (per unit land)
than rain-fed agriculture, since the farmer can control the amount and
timing of water delivery rather than being dependent on the vagaries of
weather (though this may not be true for tail-end farmers who are at the
downstream end of a canal and receive only the water that is left over
after upstream farmers have taken their share). Thus, while irrigated land
represents only about 18% of all cropland globally, it produces about 40%
of all crops.

Perhaps the most serious problem facing irrigated agriculture is that
of salinization. When irrigation water is applied to arid soils, the water
itself ends up transpiring back into the atmosphere, but any salts that were
carried in the water end up remaining in the soil. Over time, these salts
can accumulate to levels that are harmful to the fertility of the soil. This
problem will occur more rapidly if the irrigation water is somewhat saline, but even water that is quite fresh has some dissolved solids in it that
can accumulate over time.

One solution to the salinization problem is to periodically flush salts
from the soil by applying sufficient water to percolate through the root
zone and leach salts with it. However, in soils that are poorly drained,
this can result in waterlogging of the soil, in which the root zone becomes saturated with water, decreasing plant productivity. In some cases,

waterlogging can also result in the water table rising into the root zone; this groundwater may carry with it its own burden of dissolved salts.

The waterlogging problem often leads in turn to the installation of artificial drainage to carry leaching water away from the root zone. This may cause problems in downstream ecosystems, since the drainage water is high in salts and may also contain toxic elements that have been mobilized from the soil. The poster child for the problems with artificial drainage is the Kesterson National Wildlife Refuge in California, where birds suffered severe deformities due to the toxic effects of selenium. Investigations showed that the main source of selenium to Kesterson was subsurface tile drainage installed in farms in the San Joaquin Valley; the combined effects of irrigation and drainage had unwittingly mobilized the selenium that was naturally occurring in the soils.

The (somewhat-dated) bible of salinization (Ghassemi et al. 1995) suggests that as of the late 1980s, some 20% of the world's irrigated land was severely damaged by salinization. There is also archaeological evidence that salinization contributed to the downfall of several irrigation-dependent early civilizations (Postel 1999). Current efforts to combat salinization include more efficient use of irrigation water, development of salt-tolerant varieties of plants, and dilution of irrigation water with fresher sources.

## 10.3. Water Productivity

The Green Revolution focused on increasing the yield of food produced per unit of land. However, if we are interested in agricultural water use, we need to turn to the concept of *water productivity*—food produced per unit of water—or, conversely, *water footprints*—the amount of water required to produce a unit of food (equations 3 and 4). The food part of this equation (denominator in equation 3) can be expressed variously as the production of either food (e.g., tons of edible plant material), calories, protein, or economic value (dollars), while the water factor (numerator in equation 3) can be either the amount of water delivered or the amount consumed (with the difference being seepage and return flow that can potentially be captured by downstream users). For production of fiber (e.g., cotton) rather than food, the product could be measured by mass or value. The scale of analysis can make a big difference to the calculation of water footprints: some analyses count only the amount of water delivered to a field, while others include the entire amount released from a reservoir (including seepage and evaporation along the way).

$$\text{water footprint} = \frac{\text{water delivered or water consumed}}{\substack{\text{food produced or calories produced or protein} \\ \text{produced or value produced}}} \quad \text{(equation 3)}$$

$$\text{water productivity} = \frac{1}{\text{water footprint}} \quad \text{(equation 4)}$$

How much water does it take to grow my food (and fiber[3])? How much water will be needed to feed the entire world in 20 years? What can be done to produce more food with less water?

The equation W=PDF provides a helpful framework for addressing these questions. In this equation—which can be applied at the global, country, or smaller scale—three factors contribute to the total agricultural water use (W): the population (P) that is being fed, the diet (D) that the population is consuming, and the water footprint (F) of that diet. Population issues are certainly important in both predicting and reducing agricultural water needs, but we focus here on the last two factors.

## Diet

The diet factor can be divided into two subfactors: the *number* of calories consumed and the *type* of food eaten. Clearly the more food you eat, the more water it takes to grow that food. It may be less obvious that the kind of food you eat has an impact on your agricultural water use. Most dramatically, meat has a water footprint of ~5–15 m$^3$ of water consumed per 1000 calories produced, while most nonmeat products have water footprints of <1 m$^3$ per 1000 calories (Falkenmark and Rockstrom 2004, CAWMA 2007). This is because of the large loss of energy (and imbedded water) in converting feed into beef. The combination of more calories and more meat means that Western diets are considerably more water intensive than other diets. For example, the average North American has a total agricultural water footprint of 1800 m$^3$/year, compared to 690 m$^3$/year for the average sub-Saharan African (Falkenmark and Rockstrom 2004).

Even within nonmeat crops, there can be significant differences in water footprints. However, the differences in reported values for a given crop are generally much higher than differences between crops. For example,

---

3. Most of the discussion below applies to fiber products, such as cotton, as well as to food products. For simplicity, we focus on food.

water footprints (m³/ton) for wheat have been reported as 780–2600 (Falkenmark and Rockstrom 2004) or 830–5000 (CAWMA 2007), while comparable values for rice are 900–1400 (Falkenmark and Rockstrom) or 620–6600 (CAWMA), despite rice's reputation as a thirsty crop. Even differences in water footprints between C3 plants and the physiologically more water-efficient C4 plants[4] are relatively small, in part because C4 crops tend to be grown in tropical climates that are more water demanding. This indicates that reducing agricultural water use is less about influencing diet choices—with the important exceptions of reducing meat consumption and overeating—and more about improving the way that crops are grown, in order to move toward the lower end of the water footprint range for each crop (see next section).

In comparing the water implications of different crops, it is important to note that the water footprint is not the same as the *crop water requirement*, which is the amount of water that the crop needs in order to grow, generally expressed in millimeters (each mm corresponds to 10 m³ of water per ha). Some crops, such as sugarcane, have high water requirements (~2000 mm, or 20,000 m³/ha) but low water footprints (~200 m³/ton), simply because their yield per unit of land is so high (~100 ton/ha). Despite its low water footprint, sugarcane is not an appropriate crop for arid areas because of its high water requirement.

## Footprints and Efficiency

As noted above, water footprints for a given crop can vary dramatically. How can we decrease water footprints (increase water productivity)? Before we can address this question, we need to define water use efficiency.

Efficiency is a key concept in agricultural water use, but unfortunately one that has a confusing array of different definitions (Jensen 2007). These differences are not just technical; they reflect different understandings of the way to move forward in improving agricultural water use. The classical definition of efficiency at the field or project scale is the ratio of crop irrigation water use (crop ET minus the amount of that ET that is derived from rain) to the amount of water diverted (equation 5). By this definition, there are two sources of inefficiency, that is, two categories of

---

4. C4 plants carry out photosynthesis differently than C3 plants. By physically separating the capture of $CO_2$ from the incorporation of that $CO_2$ into a usable molecule, C4 plants are able to draw $CO_2$ more efficiently from air. This means that they need to open their stomates less often, resulting in less loss of water per unit of photosynthesis.

water that are diverted but don't go to crop ET: nonbeneficial ET (e.g., evaporation from a canal, transpiration from noncrop plants) and return flows (surface return flows and seepage).

The idea that return flows are wasteful makes sense if what you care about is the narrow issue of delivering the "right" amount of water to a field. But if you care about increasing the water productivity of an entire basin, it is important to note that in many cases return flows are not really wasted—they are being used downstream—and reducing them will not increase basin water productivity. As early as 1977, Jensen introduced the concept of *net efficiency*, which took into account the availability of return flows for reuse. Keller and Keller (1995) modified this to *effective efficiency* by also accounting for changes in the water quality of return flows (equation 6). Effective efficiency is the ratio of crop irrigation water use to the amount of water *depleted*, rather than to the amount of water *diverted*, where water depletion includes ET (but not return flows) and also includes a factor related to the increase in salinity of the return flow relative to the diverted water.

$$\text{classical efficiency} = \frac{\text{crop ET} - P_e}{\text{DIV}} \qquad \text{(equation 5)}$$

where

  $P_e$ = effective precipitation (the amount of rainfall that is available for crop ET)
  DIV = total amount of water diverted for delivery to the study area

$$\text{effective efficiency} = \frac{\text{crop ET} - P_e}{\text{DIV} - R_e} \qquad \text{(equation 6)}$$

where

  $RF_e$ = effective amount of return flow (accounting for degradation of water quality)

With these definitional issues behind us, we can discuss some measures that can be used to increase agricultural water use efficiency.

*Switching to improved irrigation technologies.* There are vast differences in efficiency between different methods of applying water to a field. Traditional *gravity irrigation*—in which water is flooded across an entire field or flows into furrows between rows—leads to high evaporative and seepage losses, with classical efficiencies of only 30–50%. *Sprinkler irrigation* can reach efficiencies of up to 85%, depending on the size of the droplets formed. Large-scale sprinkler irrigation is usually carried out with center-pivot systems, in which large circular fields (~50 ha) are irrigated with computer-controlled spray nozzles that rotate around a central pivot point. *Drip irrigation* consists of water pipes (surface or subsurface) with emitters—placed directly where plants are located—that release a continuous slow drip of water to the plants. This achieves high water delivery efficiencies (>90%) and also typically results in higher yields, since water is continually available to the plants (Box 10.1). The installation cost of conventional drip irrigation (>$1000 per ha) is prohibitive for farmers in many developing countries, but lower-tech, low-cost drip systems have been developed that have payback periods of less than a year even for poor, small-scale farmers. There are also other methods that can make any irrigation technology more efficient, such as precision leveling of fields (so that water is applied uniformly) and providing timely information on crop water requirements.

*Lining irrigation canals.* When water is transported long distances (e.g., from a dam to a far-off irrigation project), a great deal of water can be lost to evaporation and seepage en route. It is difficult to do much about evaporation (except for using closed systems, which is often prohibitively expensive), but there have been efforts in many places to reduce seepage by lining the bottoms of canals with an impermeable material. However, since seepage is not a consumptive use, the effectiveness of this measure depends on which definition of efficiency you accept (above) and, in particular, on whether the seepage water is being utilized by other farmers. A prime example of the unforeseen consequences of canal lining is the All-American Canal, which carries Colorado River water to the Imperial Irrigation District in California. The $200 million effort to line 23 miles of this canal is expected to "save" ~86 MCM of water annually, which will then be sold to San Diego for household use. However, there is evidence that a significant portion of this seepage has been recharging groundwater in the Mexicali Valley in Mexico, where farmers who have been using it for their own irrigation will be the losers. In short, canal lining improves classical efficiency but not effective efficiency. The canal lining project has led to cross-border tensions for years, including an unsuccessful lawsuit.

*Increasing yields.* Farms with low yields (per unit land) typically have low water productivities. In these farms, increases in crop yields generally lead to only small increases in water consumption but large increases in

## Box 10.1. Modeling the Effects of Drip Irrigation Subsidies in the Upper Rio Grande Basin

The Rio Grande originates in Colorado, flows through New Mexico, and serves as the Texas-Mexico border before draining into the Gulf of Mexico. The waters of the river are fully allocated for drinking water and irrigation, and the basin is currently suffering from the multiyear southwestern drought. Ward and Pulido-Velazquez (2008) constructed an integrated hydrologic-agricultural-economic model for the upper basin and used it to analyze how subsidies for drip irrigation would affect land and water use in the basin. They predict that if government-offered subsidies were to be increased from 0 to 100% of drip capital costs, the amount of land under drip irrigation would increase sevenfold, mostly due to conversion from flood irrigation but also due to an increase in total land under irrigation. Under the high-subsidy scenario, the amount of water applied to fields would decrease from 364,000 to 324,000 acre-feet/year. However, due to the greater efficiency (classical definition) of drip irrigation, more of the applied water would be consumed, resulting in an *increase* in consumptive use from 167,000 to 204,000 acre-feet/year. Based on this, the authors reach the conclusion that "water conservation in irrigation can increase water use."

I believe that this conclusion is somewhat misleading. Drip irrigation is highly efficient by both the classical and effective efficiency definitions. By applying water only where it is needed, drip irrigation doesn't just reduce return flows (which is arguably not an efficiency improvement) but also reduces nonbeneficial ET (which indisputably *is* an efficiency improvement). The bottom line is that drip irrigation increases water productivity and reduces water footprints. The reason that Ward and Pulido-Velazquez's (2008) model shows greater consumptive use in the Rio Grande under a higher-drip scenario is simply that more food is produced. While this scenario may be a problem for other users in the basin, it would be a good thing for total global agricultural water use.

I agree with Ward and Pulido-Velazquez (2008) that it is crucial to examine the effects of different irrigation technologies on consumptive water use, not just water withdrawals. However, I believe that the statement "water conservation in irrigation increases water use" will be true only if more food is being grown and that water conservation measures such as drip irrigation make a valuable contribution by allowing us to grow more food with lower water withdrawals and with lower water consumption per unit of food grown.

food production, resulting in a net decrease in the water footprint. Measures to increase yields include soil improvement, application of fertilizers, and utilization of HYVs. There is great potential to improve water productivity in parts of the world that have not yet fully realized the benefits of the Green Revolution.

*Using supplemental irrigation.* Improving water efficiency means making better use of rain-fed agriculture as well as irrigated areas. One way to improve the water productivity of rain-fed agriculture is to provide supplemental irrigation. This can help avoid the decreases in yield—or even crop failures—that can result from periods of dry weather.

*Becoming aware of natural productivity.* When water is used for agriculture, it is being taken from another user—often the environment. Before diverting that water to agriculture, it is important to be aware of the ecosystem services that it is providing, which in some cases are in themselves providing a food source—fish, nuts, wildlife—that will be lost if the water is diverted.

*Removing water subsidies.* Water for irrigation is often provided at prices that are ridiculously low compared to the true value of the water and compared to the prices being paid by industrial and domestic users. A more reasonable pricing structure would encourage more efficient use of this vital resource. It would also allow us to better distinguish between those irrigation projects that make sense and those that don't, and would drive agriculture out of locations where higher-valued users need the water. Ultimately, charging farmers for the true value of the water they use would probably mean higher food prices, which would be a good thing for the long-term sustainability of agriculture.

We can summarize our discussion of agricultural water use by returning to the equation $W = PDF$. I have modeled this equation on the more general $I=PAT$ equation, which sees environmental impact (I) as a function of population (P), affluence (A), and technology (T). Like IPAT, WPDF can be used either qualitatively, as a way of describing the factors determining water use, or quantitatively, to calculate agricultural water use for a given area. For quantitative use, it would take the form shown in equation 7, in which the amount of each type of food consumed is multiplied by the water productivity of that food. In any case, our task is to tackle all three factors—population, diet, and water productivity—in order to minimize the total water required to provide all of Earth's people with enough to eat.

$$W = P \sum D_i F_i$$

units:

$$m^3 = people \sum \frac{calories}{person} \frac{m^3}{calorie}$$

(equation 7)

where

D$_i$ = average calories consumed of an individual food
F$_i$ = water footprint of that food

## 10.4. Agricultural Pollution

The amount of land used for agriculture worldwide (including both crops and grazing) is now relatively stable at around 36% of the world's land area, excluding Antarctica (FAOSTAT). Thus, agriculture is a globally important land user as well as a direct water user. As a result, agriculture can contribute pollution in two ways (which are sometimes hard to distinguish): through return flow of polluted irrigation water and through runoff from agricultural lands. As such, agricultural pollution can walk a fine line between point and nonpoint source pollution.

As noted in Table 8.4, agriculture is the top-ranked source of pollution to US rivers. Some of the most important pollution problems from agriculture include the following:

- Nutrients: Large quantities of nitrogen (N) and phosphorus (P) can be delivered from fields either as subsurface flow or as surface runoff, especially in areas where chemical fertilizers are applied at high rates and where flow pathways to streams have been shortened (e.g., through artificial drainage or removal of vegetated buffers around streams). Studies in the US suggest that only 14% of fertilizer N applied to crops eventually reaches humans through nitrogen in food, with the remainder being released to surface water, volatilized into air (as the noxious compound ammonia), stored in soils, or lost to denitrification (NRC 2000). On the way to these ultimate fates, much of the fertilizer N ends up in the form of animal manure,[5] which is often stored in ponds that are susceptible to releasing large volumes of waste during storm events. Attempts to close the nutrient loop

---

5. The EPA estimates that the total amount of manure generated in the US is more than three times the amount of human waste.

by applying manure to fields (instead of using chemical fertilizer) have foundered on the fact that industrial agriculture has increasingly segregated animals and crops into different parts of the country.

- Sediment: Bare soils are susceptible to erosion, which represents both a loss of productive topsoil and a water quality issue for receiving waters. Sediment can also carry with it many pollutants, including P, metals, and organic contaminants. No-till agriculture—in which the soil is not plowed to prepare it for seeding—has gained popularity, in part as a way of reducing erosion, though it often requires increased use of herbicides to kill weeds.

- Pesticides: The agricultural sector is responsible for the bulk of the 1 billion pounds of herbicides, insecticides, fungicides, and fumigants that are used each year in the US alone. While newer generations of pesticides tend to have reduced effects on non-target species (including both humans and wildlife), there are still significant environmental and human health risks associated with these compounds. In addition, many compounds that have been banned in industrialized countries are still in use, legally or illegally, in developing countries.

- Indicator bacteria: Manure potentially contains numerous pathogens, which can be carried into groundwater or surface water, especially when excessive volumes of manure are applied to fields or stored in lagoons.

Because of the political power of agriculture and the complexity of agricultural runoff, the regulation of agricultural pollution in most countries has lagged behind that of industrial and municipal point sources. A recent focal point of agricultural regulation in the US has been *concentrated animal feeding operations* (CAFOs), where large numbers of animals are grown in confined feedlots. CAFOs generate huge quantities of manure that often exceed the capacity of nearby fields to absorb. As a result, they can be significant sources of nutrients and pathogens to wells and rivers. The Clean Water Act treats CAFOs as point sources requiring a permit to discharge, but there has been a lengthy struggle in the courts and at EPA to define which CAFOs are covered, what counts as a "discharge," and what requirements should be contained in the permits.

## 10.5. Conclusion

In this chapter, we have discussed issues related to the use of water and land for growing food and fiber, including the successes and shortcomings of the Green Revolution, ways to evaluate and improve water use efficiency and water productivity, and the nature of agricultural pollution.

The sheer volume of agricultural water use means that this sector is a huge contributor to our current water scarcity crisis. When we look 30 or 50 years into the future and imagine trying to feed a world of 9 or 10 billion people—including satisfying the demand for more meat in rapidly developing countries while also reducing the current, unacceptable extent of malnutrition in the poorest countries—it can be easy to see a world in which water is truly stretched beyond its limits. Yet, ironically, the current patterns of wasteful water use (often encouraged by perverse subsidies) provide some hope. Is there enough inefficiency in the agricultural water system that we can increase food production dramatically without using much more water? Can the measures outlined in section 10.3 allow us to solve the W=PDF equation while still preserving water for other human and ecosystem needs? Can a reduction in agricultural pollution help free up clean water for human use and the restoration of aquatic ecosystems?

We turn now to the third main sector of water use (besides agriculture and households): industry.

# 11

## Growth and Sustainability: Using Water More Wisely in Industry

Industrial activity has been a key driver of economic growth and improved standards of living since the Industrial Revolution. Over the last several decades, however, we have increasingly bumped up against the limitations of that model and have been confronted with a new set of questions: Can we continue growing indefinitely in a world of finite resources? How do we balance growth with environmental and social sustainability? How do we move toward improved quality of life rather than simply increased quantity of economic activity? How do we ensure that poorer countries have a chance to develop their economies without experiencing the same resource depletion and environmental degradation that have accompanied growth in rich countries?

These questions go well beyond the scope of this book and are addressed by several excellent recent books by prominent thinkers (McKibben 2007, Speth 2008, Sachs 2009). Here we limit ourselves to a narrower, but related, set of questions: How can we move industrial activity toward a more sustainable model of water use? What are the ways that industry currently affects water resources, and how can we minimize any negative impacts? The first section deals with the energy sector and explores the many links between energy and water, while the second section deals with water impacts of other industrial sectors.

## 11.1. Water and Energy

Industrial economies are massive users of energy. All steps in energy use, from mining to electricity generation, have potential impacts on water resources. At the same time, water management has implications for energy use. Over the last several years, there has been increased attention to the multiple linkages between these two scarce resources (e.g., DOE 2006, Webber 2008, Sehlke 2009). This section explores those linkages.

### Water for Energy: Fossil Fuel Extraction

Extraction of coal, oil, or natural gas from the ground is the first step in generating most of the energy we use today. Although the mining industry uses a relatively small amount of water (Figure 3.12), fossil fuel extraction can have serious pollution impacts. Specific issues have arisen around each of the main fossil fuels.

Petroleum. The extraction and transportation of petroleum around the world can lead to release of large volumes of oil into the environment, where it can have harmful effects on aquatic ecosystems and on the safety of water resources. Catastrophic spills, such as the *Exxon Valdez* spill in Prince William Sound, Alaska, can create short-term devastation and long-term repercussions. A National Research Council study (NRC 2003) found that spills represented less than 10% of annual global petroleum releases into the environment, with the majority of the releases coming instead from chronic, low-level sources, such as land runoff (from use of petroleum products) and natural seeps. Gasoline additives can also pose environmental hazards, with the contamination of groundwater wells by MTBE (methyl *tert*-butyl ether, an additive used to reduce air pollution) being a prominent example.

Coal. Perhaps the greatest damage from coal extraction comes from the practice of "mountaintop removal mining," in which ridgetops are dynamited in order to get at the coal underneath. The rock and debris that are removed are dumped in valley bottoms, where they smother streams and contaminate local water sources. This practice has been around since the 1960s but has accelerated in recent years, especially in Kentucky and West Virginia, as mining companies have raced to extract low-sulfur coal made more desirable by clean air regulations. Various legal efforts have been made to restrict mountaintop removal mining, but for now the practice continues unabated.

Natural gas. Technological advances and the desire to reduce greenhouse gas emissions have led to a boom in natural gas extraction in the US. In particular, two previously uneconomic gas sources are now being tapped: shale gas and coalbed methane.

*Shale gas* refers to natural gas (mostly methane) that is trapped at relatively low concentrations in shale deposits. This gas is now extractable, thanks to two new developments: horizontal drilling techniques, which can vastly increase the area of the well that is in contact with the shale gas; and hydraulic fracturing ("fracing"), in which high-pressure water, along with chemicals and sand, is used to create fractures in the rock through which the gas can travel. There are many anecdotal reports of contamination of groundwater with natural gas or with fracing chemicals (Mouawad and Krauss 2009), although an EPA study concluded that there was no evidence that fracing contributes to groundwater contamination (EPA 2004). Shale gas is found in various parts of the US, but some of the most active drilling is taking place in the Marcellus Shale, which extends under large swaths of New York, Pennsylvania, Ohio, and West Virginia. Under pressure from government officials and environmental groups, Chesapeake Energy Corporation, one of the main players in shale gas extraction, recently announced that it would not drill in the part of the Marcellus Shale that underlies the watershed for New York City's water.

*Coalbed methane* (CBM) refers to methane that is trapped in coal deposits. Extraction of this gas involves first depressurizing the system by pumping out water, generally for several months, before gas production can begin. Hydraulic fracturing is also used in order to increase the flow of gas. The center of CBM production in the US is the Powder River Basin in Montana and Wyoming, where, as of 2006, there were about 20,000 wells operating, with an additional 2000 wells being added per year (Swindell 2007); each well discharges about 18 m³ of water per day (Colmenares and Zoback 2007). Disposal of this water is becoming an increasingly serious problem, especially since some of it is contaminated or brackish. In addition, the removal of large volumes of water in this semiarid area may lead to dropping groundwater levels and increasing difficulty in obtaining water for irrigation and household uses.

## Water for Energy: Thermoelectric Power Plants

As noted in Chapter 3, thermoelectric power plants in the US have greater water withdrawals than any other sector, although their consumptive fraction is quite low. Data for other countries are not readily available, since AQUASTAT does not separate power plants from other industrial uses. It is also likely that not all power plant withdrawals are reported to AQUASTAT, since these withdrawals are often ignored, given that the water is usually taken directly from a surface water source and returned to the same source with minimal consumptive use. Still, the large volumes of water involved suggest that efforts should be made to improve reporting.

The high water requirements of energy generation play an important role in determining where power plants can be built. As water scarcity becomes more prevalent in the US, water availability is becoming more and more of a constraint in siting new power plants. As Wolfe et al. (2009) note, "Opposition to power plant siting frequently focuses on water use, especially (but not exclusively) in water-short regions." Even existing plants in some locations are in increasing danger of having to shut down during low water conditions (Webber 2008).

Power plants use water primarily for cooling of the steam formed during electricity generation. Older power plants often use "once-through cooling," in which water is withdrawn from a river or lake, directed through a heat exchanger (in which it cools the steam and is itself heated by several degrees), and discharged back to the environment. The combination of increasing water scarcity and more stringent environmental regulation[1] has led to an increase in the use of "recirculating cooling," which involves using the same parcel of cooling water repeatedly. The heat transferred to the recirculated water is dissipated by evaporation of some portion of the water, either in a cooling tower or in an on-site pond. The amount of new water ("make-up water") required in recirculating cooling is determined by the evaporation rate and by the frequency with which water must be drawn off and discharged ("draw-off water") in order to prevent the concentration of dissolved solids from getting too high. The current distribution of water cooling systems in the US is about 43% once-through cooling, 42% cooling towers, 14% cooling ponds, and 1% air-cooling (Feeley et al. 2008).

The shift from once-through to recirculating cooling has led to a dramatic decline in power plant water withdrawals in the US over the last 30 years (Figure 3.13), even as electricity generation has continued to grow.

---

1. Section 316 of the Clean Water Act required regulation of both heat discharges and water intake structures in order to better protect aquatic life from temperature modification and from impacts associated with entrainment of organisms into the water intake.

The average water withdrawal per unit of energy production decreased from 63 gallons per kilowatt-hour (gal/kWh) in 1950 to 23 gal/kWh in 2005 (Kenny et al. 2009). However, this decline appears to be leveling off (Figure 3.13). In addition, there is growing concern over the fact that re-circulating cooling, despite its much lower water withdrawals, actually has higher consumptive water use (Feeley et al. 2008). Dziegielewski and Bik (2006) calculate that the average fossil fuel plant with once-through cool-ing consumes 0.2 gal/kWh, while recirculating systems consume 0.7 gal/kWh; numbers are slightly higher for nuclear plants. Feeley et al. (2008) give slightly different numbers for each of these categories, but show the same pattern. Thus, in some basins, the shift to recirculating cooling may actually result in insufficient water for other users or for environmental flows. Efforts are under way at the Department of Energy and in the electricity industry to develop ways to reduce consumptive water use.

The increase in the number of combined-cycle power plants is also helping reduce water use in power generation. These plants use the tra-ditional steam turbine to generate only about a third of their power, with a gas turbine supplying the remainder. This automatically reduces cool-ing water use (both withdrawal and consumption) by about two-thirds (EPRI 2002).

An additional issue with power plants, especially coal-burning ones, has to do with water pollution (Duhigg 2009b). As air pollution regula-tions have become more stringent, more and more plants have installed scrubbers to capture pollutants before they escape the stacks. In many cases, these scrubbers generate a new stream of wastewater, which is poorly regulated and often hazardous.

## Water for Energy: Hydropower

Water can also be used to directly generate electricity, by using the power of falling water to drive a turbine. The energy produced is proportional to the flow of water and to the *head*, the elevation difference between the water surface and the turbine. Hydropower generally involves construct-ing a dam in order to increase the head and smooth out the temporal variability in flow. Some hydropower plants (including the second-largest hydropower producer in the US, the Chief Joseph Dam on the Colum-bia River, with an installed capacity of 2620 MW) are "run-of-river" projects, which have no capacity to store water, though they often still include a dam in order to build up the head. Run-of-river projects are most common in rivers with fairly constant flow or with upstream dams that can regulate flow.

There is currently a push toward projects that can generate small amounts of power without requiring a dam. Hall et al. (2006) estimate that 5400 streams in the US could feasibly[2] be developed for "small hydro" (1–30 MW)[3] and could collectively produce as much as 18,000 MW. An additional ~120,000 sites could produce ~12,000 MW of "low power" hydro (<1 MW). This compares to current US hydropower production of ~35,000 MW, of which ~20% is already produced by small hydro and low power sites, though not all of these are dam-free sites. The environmental impact of the extensive small-scale hydropower development envisioned by Hall et al. (2006) has not been sufficiently evaluated and is likely to be substantial. Still smaller hydropower projects that are unlikely to have significant environmental impacts (microhydro: <2 kW; nanohydro: <100W) are being developed and used on a grassroots level for powering individual households.

Hydropower currently produces 2.2% of world energy use and 19% of electricity use. The extent of hydropower development varies around the world, with North America, OECD Europe, China, and Brazil having the greatest installed capacity. Only North America, OECD Europe, and Australia have tapped more than 50% of their economically exploitable hydropower potential, and hydropower is projected to grow globally by almost 2% per year for the next several decades (WWAP 2009).

As noted in Chapter 3, hydropower in the US "uses" 10 times more water than all other users combined. This instream use is really not directly comparable to other uses. Still, there are certainly environmental impacts associated with this use, as well as a consumptive component—reservoir evaporation, which consumes more water globally than industrial and household uses combined. Of course, it is no easy task to determine how much of that reservoir evaporation is attributable to hydropower operations, as opposed to other dam functions.

Hydropower production is, of course, dependent on the flow of water. At low flows, hydropower plants will not be able to operate to their full capacity, while at high flows, they may need to release excess water through spillways without using it to generate electricity. In basins where high water

---

2. Feasibility is defined here to include limitations based on the following criteria: site access, proximity to transmission lines, environmental sensitivity, the amount of flow that would be allowed to be diverted for power production (up to half of streamflow), and the permitted maximum length of the penstock through which the water would flow (300 to 15,000 feet, depending on region).

3. Different organizations use different definitions of small hydro, low power hydro, microhydro, and nanohydro.

consumption by other users leads to reduced streamflows, downstream hydropower installations will have decreased power production.

Dam advocates are quick to point out that hydropower is currently the largest source of renewable energy and argue that it represents an important tool in the battle against climate change. Indeed, most people assume that hydropower does not produce any greenhouse gases, in contrast to the vast quantities of $CO_2$ produced by fossil fuel combustion.

In reality, however, reservoirs—whether for hydropower or other purposes—do release both $CO_2$ and the more potent greenhouse gas $CH_4$ (methane). These compounds originate from the decomposition of organic matter within the reservoir. Sources of organic matter include particulate matter washed in from upstream, algae produced in the reservoir, and trees that occupied the area before flooding. A rough estimate suggests that the world's reservoirs produce about 7% of the world's total greenhouse gas (GHG) emissions (St. Louis et al. 2000).

How do GHG emissions from hydropower compare to those from fossil fuel plants of similar capacity? This question has proven both scientifically difficult and politically controversial. What has become clear is that there is a large variation in hydropower GHG emissions per unit of electricity produced. Factors that tend to lead to relatively high GHG emissions include the following:

- Low power intensity: Power intensity ($W/m^2$) is a measure of how much power is produced per unit of land flooded. Since GHGs are emitted from the entire surface of the reservoir, shallow reservoirs that have a high surface area but produce little power are particularly likely to have relatively high GHG emissions.
- Tropical location: Tropical reservoirs have much higher GHG emissions than temperate ones, due to the larger amounts of organic matter present and the higher rates of decomposition.
- Subsurface water release: Relative to surface waters, deeper waters tend to be high in $CH_4$, which is produced in the sediments and gets oxidized to $CO_2$ on its way up the water column. When dams release bottom water, the oxidation process is bypassed and large amounts of $CH_4$ are released to the surface and then to the atmosphere. Since each molecule of $CH_4$ has as much global warming potential as about 25 molecules of $CO_2$, this can have a huge impact.

The worst sites—those that share all three of these characteristics, such as the Petit Saut reservoir in French Guiana (Delmas et al. 2001) or the

Samuel project in Brazil (Fearnside 2005, dos Santos 2006)—have GHG emissions that can be considerably higher than a comparable natural gas power plant, at least in the short term.

## Water for Energy: Bioenergy

There has been increasing interest recently in using biomass (plant material) as a source of renewable, domestic energy. Biomass can generate energy in two ways: by combustion in a power plant to produce electricity or by conversion to a liquid fuel (biodiesel, bioethanol) that can be used to power vehicles. At present in the US, the main action involves biofuels, primarily ethanol made from corn, although next-generation biofuels will rely on other (nonfood) crops.

What are the implications of increased biomass production for water scarcity and water pollution? Clearly the consumptive use of water (blue or green) for growing crops is much higher than for extracting fossil fuels, so relying on agriculture to produce energy as well as food could lead to a large increase in water consumption, exacerbating water scarcity. An analysis of the water footprints of different forms of bioenergy (Gerbens-Leenes et al. 2009) found that bioelectricity had a lower footprint (m³ of water per gigajoule of energy) than biofuels, simply because more of the crop can be used. For ethanol production, crops with relatively low water footprints (<150 m³/GJ) included sugar beet, potato, sugarcane, maize (corn), and cassava.

Water quality is also likely to be negatively affected by increased bioenergy production. Agriculture is already the biggest polluter in the US (see sections 8.7 and 10.4), and increasing the area devoted to corn and other crops is likely to lead to an increase in fertilizer and pesticide runoff. A National Research Council synthesis (NRC 2008) notes, "The growth of biofuels in the United States has probably already affected water quality because of the large amounts of N and P required to grow corn." Many people are hoping that technology can solve the problem—in particular, that newer-generation biofuels will have lower fertilizer and water requirements and will be grown primarily on marginal land using green water instead of displacing food crops from the limited supply of arable land and irrigation water.

## Energy for Water

Until now, we have dealt with the impacts of energy use on water resources. We close this section with a brief examination of the opposite issue—the impacts of water management on energy use.

All steps in the water use cycle require energy: pumping water from the source (surface or groundwater); treating it; delivering it to homes, businesses, and farms; and collecting and treating the resulting wastewater. The Department of Energy (2006) estimates that 4% of total US electricity use goes to power the water cycle. Certain types of water management activities—long-distance water transport and desalination, for example—are much more energy-intensive than others.

Not included in the DOE estimate is energy consumed by the end user (e.g., home or business) to heat, cool, filter, or pump water. A California study (CEC 2005) estimated that 14% of the state's electricity use and 31% of its natural gas use is related to those end uses. However, it is not clear that this energy use should truly be attributed to the water sector, since it includes the energy used in (water-based) home heating and cooling as well as the energy used for true water uses, such as heating of water used by dishwashers. The California example does help point out that water conservation in all sectors can reduce energy use, by reducing the need to heat or cool that water as well as by reducing the energy needs for supply, conveyance, and treatment.

## 11.2. Water and Industrial Production

We now turn away from the energy sector to examine the use of water in other industrial sectors, including manufacturing and forestry.

### Improving Efficiency

All types of industrial activity require some water. Because of the complexity of industrial cycles of materials and energy, it can be extremely difficult to figure out exactly how much water is imbedded in a given industrial product or service. Nonetheless, some estimates do exist. Hoekstra and Chapagain (2007) use their footprint method to estimate that producing a sheet of paper requires 10 liters of water, while producing a cotton T-shirt requires 2000 liters (see Table 3.3).[4] McCormack et al. (2007) conducted a sophisticated input-output analysis of 17 different buildings in Australia and found that the water imbedded in the

---

4. These estimates (and Hoekstra and Chapagain's footprint method in general) are meant to include the total amount of blue or green water consumed or polluted along the entire supply chain. However, they are in some sense underestimates, since they don't include the water footprint of the energy used in growing or processing these products (although the *consumptive* use associated with that energy use may be rather small).

construction of these buildings ranged from ~5000 to 20,000 liters per m² of gross floor area.

In order to deal with water scarcity while still allowing for improved standards of living, especially in developing countries, there is an urgent need to improve the efficiency of industrial water use. There is substantial evidence that this can be done. For example, Figure 11.1 shows industrial water withdrawals in the US (1950–2005), along with the Total Industrial Production Index. It is clear that industrial production in the US grew substantially since 1950, while the water necessary to fuel that production has declined. The water used per unit of industrial production dropped by a factor of 7, although this decline leveled off in recent years.

The techniques used to achieve increased efficiency will vary tremendously from one industrial sector to another, making it hard to come up with cross-sectoral recommendations or regulations. However, some basic tools can be used to encourage innovation in reducing water use: requiring the metering and reporting of water use, raising water prices, and stricter regulation of industrial discharges.

## Water Risks and Opportunities

Companies in many sectors and countries are starting to realize that, like energy, water is a scarce resource that is essential to their productivity and their bottom line. Some companies are even developing indices or other ways of assessing water-related risks to their operations. These risks can include insufficient water quantity (e.g., due to drought, climate change, or an overallocated river), inadequate water quality (due to chronic or episodic contamination), a restrictive regulatory environment (requiring large investment in water treatment), and unfavorable media reports or community opposition (e.g., over pollution attributed to company operations). To mitigate this last risk, reporting on water management is increasingly part of corporate social responsibility (CSR) reports (Morikawa et al. 2009).

At the same time, some companies are seeing water scarcity and pollution as business opportunities and are developing and marketing new technologies, especially for water treatment. Globally, the water treatment business is worth $300–$350 billion per year (Voith 2008) and includes a variety of traditional and innovative approaches.

## Industrial Pollution

Effluent from industrial facilities can contain a tremendous variety of pollutants, including substances that are toxic to people and aquatic

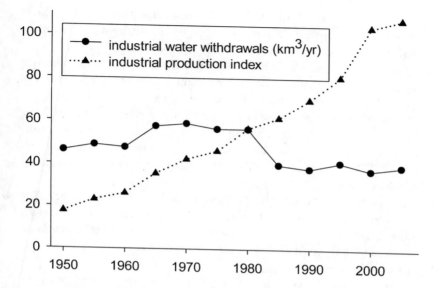

*Figure 11.1.* US industrial water use and industrial productivity, 1950–2005. Water use data from Kenny et al. (2009), including mining and aquaculture, but excluding salt water (see caption to Figure 3.13). Total Industrial Production Index (2002 = 100) from Federal Reserve (www.federalreserve.gov/datadownload/Choose. aspx?rel=G.17).

organisms. When this effluent is dumped untreated into rivers or other water bodies, it can contaminate large volumes of water, making it unfit for human use or for supporting life. This was the case in the US not so long ago, when rivers would catch fire from oil releases or run different colors depending on the dyes being used by the local factory. The Clean Water Act changed all that, by requiring industry to treat its effluent before discharging it under a NPDES permit. Despite the enforcement problems noted in Chapter 2, the Clean Water Act has without a doubt reduced industrial pollution significantly.

Two consequences of the NPDES system are worth noting. First, the more stringent discharge requirements helped contribute to the substantial decrease in industrial water use noted above, since using less water means having less effluent to treat. Second, a large number of smaller facilities no longer discharge directly to surface water but instead discharge to a municipal sewage treatment plant. While this provides the benefits of centralized waste treatment, it also brings with it the risk that industrial waste will contain substances that are not effectively treated by the

STP—or, even worse, are toxic to the biological system that is the heart of secondary treatment. "Pretreatment" requirements are meant to ensure that these risks are minimized.

In addition to the Clean Water Act, there is another important piece of federal legislation that has contributed to the reduction in industrial pollution. The Emergency Planning and Community Right-to-Know Act of 1986 (EPCRA) was a response to the 1984 accident in Bhopal, India (in which toxic gases released from a Union Carbide facility killed thousands of people), and a smaller accident at a similar facility in West Virginia. EPCRA established the Toxics Release Inventory (TRI), a database of self-reported industrial use and release of toxic chemicals. The exact rules governing TRI have changed several times over the years, including significant changes to the chemicals that are covered (now numbering more than 600), the types of facilities that must report to TRI (now more than 21,000 individual facilities), and the categories of chemical transfers that must be reported.

One difficulty in working with TRI data is that they are generally reported as the total pounds of chemicals that are released or disposed of in a certain way (e.g., discharged to surface water). Given the widely varying toxicity of these chemicals, a better understanding of risk requires looking at individual chemicals or scaling them by their toxicity.

The logic of TRI was that arming communities with this knowledge would result in public pressure to reduce the use and release of toxic chemicals. In this sense, the program has arguably been quite successful. EPA and several environmental groups make the data widely available and encourage the public to locate the polluters in their neighborhoods. In addition, the reported total release/disposal of toxic chemicals has dropped significantly, as shown in Figure 11.2 (top). However, examining the data on discharges to surface water (Figure 11.2 bottom) shows a different pattern: discharges of the original (1988) chemical list have dropped, but the (much larger) discharges of the 1995 chemical lists have increased. At the same time, there have been questions raised as to whether the public really tracks TRI data (Atlas 2007) and whether the reported declines represent real reductions in risk as opposed to simply changes in reporting methods or shifts to chemicals that are equally toxic but are not yet on the list (Natan and Miller 1998).

## Forestry

One industrial sector that can have a large impact—positive or negative—on water resources is forestry. Forests that are managed for timber occupy

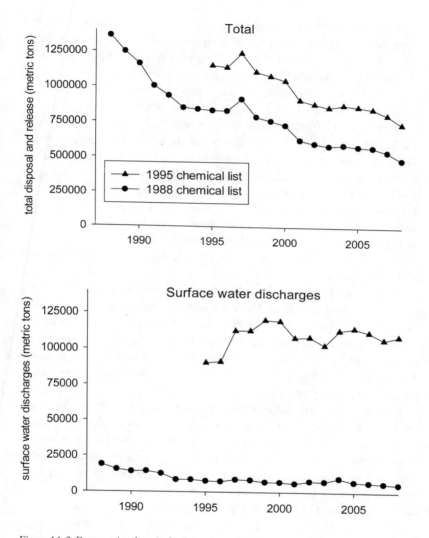

*Figure 11.2.* Patterns in chemical release over time. Top: Total on-site and off-site release and disposal. Bottom: On-site surface water discharges. Note the 10-fold difference in vertical scale between the top and bottom graphs. Each curve shows one consistent set of chemicals over time (either the chemical list as of 1988 or the chemical list as of 1995). Data source: TRI (www.epa.gov/triexplorer/).

large expanses of land throughout the world. Increasingly, timber management is being supplemented with managing for other forest products, such as nuts or mushrooms or biodiversity. Or water. Mike Dombeck, former chief of the US Forest Service, points out that water is one of the most important, but underappreciated, products of the National Forests, with a value estimated at $3.7 billion per year (Dombeck 2003).

How should forests be managed for water? As discussed in section 6.2, cutting trees tends to increase water quantity but decrease quality. The goal of watershed forestry is to achieve high water yields, while also maintaining the ability of the forest to serve its protective hydrological functions: extensive storage of water in the soil, protection of the soil from erosion, and slow subsurface movement of precipitation toward the stream or reservoir, with associated filtration and retention of pollutants. These functions are best served by a mixed-age forest of intermediate density, which has higher water yields than a dense even-aged forest and is also less susceptible to disturbances, such as disease outbreaks, fire, or wind, that would leave large areas of bare, unvegetated soils. Producing such a mixed-age, resilient forest requires active management, including logging. Other goals of watershed forestry might include providing breaks in the canopy that are oriented for maximum snow accumulation, ensuring a robust litter layer and a healthy biological community in the soil, and providing special protection to hydrologically sensitive areas, such as wetlands, riparian zones, and steep slopes.

Different goals will dominate watershed forestry in different locations. In some cases—such as on water company land surrounding a reservoir in the eastern US—the primary management goal will be protection of water quality. In other cases—such as the more arid mountain West—it may be more important to increase water quantity, manage snow, and avoid devastating fires. In still other locations, the goal may be to produce the maximum amount of timber while not overly affecting water quality.

In any case, logging should be done in ways that minimize the negative hydrologic impact of the logging operations themselves. For example, forwarders—trucks that carry, rather than drag, trees to the landing—should be favored over skidders; and care should be taken in the placement of logging roads and the timing of logging operations to minimize the damage to soils in hydrologically important parts of the watershed.

## 11.3. Conclusion

Following our discussions of household and agricultural water use in the previous two chapters, this chapter covered the water quantity and quality issues associated with the industrial sector. Industry is not nearly as large a water user globally as agriculture, and it doesn't meet as basic a need as household use. Still, it is a very important player in many basins, in terms of both water use and water quality impacts.

Without water, we can't produce the energy or make the products that are so central to our economies and our lives. Making better use of industrial water is a key component of solving the water crisis and ensuring that there is water for all for both direct and indirect uses.

With an understanding of individual sectors under our belts, we now turn to two chapters that cut across sectors and address the economics, law, and politics of water allocation and better water management.

# 12

## Basic Need and Economic Good: The Contested Role of Economics in Water Management

A great battle is raging in the world of water management over the proper role for economics. On one extreme are those who see water as purely an economic good whose allocation should be determined by the marketplace. On the other are those who see water as a basic human right that should not be governed by economic considerations. Of course, these extreme positions are to some extent caricatures: there are many people in the middle who understand that water is both a basic right and an economic good and that we must find ways to manage water both equitably and efficiently.

Indeed, both perspectives on water are well established in international agreements. The existence of a basic human right to water currently derives from the right to health, which was established in the International Covenant on Economic, Social, and Cultural Rights (entered into force in 1976). In 2002, the UN committee responsible for this covenant adopted General Comment 15, which recognized that the right to health presupposes a right to "sufficient, safe, acceptable, physically accessible and affordable water for personal and domestic uses."[1] The treatment of water as an economic good also has a notable history, going back at least as far as the International Conference on Water and the Environment in Dublin in January 1992, where one of the four principles established for

---

1. There is currently a movement to enshrine the right to water as a separate, clearly articulated right of its own.

water management was that "water has an economic value in all its competing uses and should be recognized as an economic good."

Clearly the key question is how to balance these two different perspectives on water—and it is over this question that heated disputes have arisen. These disputes are playing out in several distinct arenas:

- To what extent should economic tools such as markets be used in deciding how to allocate the rights to water sources?
- Should water pricing reflect the true costs of water or the ability of consumers to pay?
- What role should private companies play in constructing and operating water and sanitation infrastructure?
- How useful is economic cost-benefit analysis in deciding which water development projects to undertake?

We discuss each of these in turn in subsequent sections of this chapter. We begin with an introduction to the economic approach to the management of resources in general and water in particular.

## 12.1. The Economic Perspective

Economics can be defined as the science of allocating scarce resources efficiently (Griffin 2006). To an economist, an *efficient* allocation is one in which the total net benefits (benefits minus costs) from the resource are maximized. In simple terms, an efficient allocation gives more of the resource to the users who value it more, in order to create the maximum total value. More precisely, in an efficient allocation, the *marginal net benefits* (MNB) for each user—the net benefits that each user gets from the next increment of consumption—are equal. This must be the case, since if the MNBs were not equal, we could achieve greater total net benefits by reallocating the resource from the user with lower MNBs to the user with higher MNBs.

Note that this theory of efficient allocation applies even when the users are consuming the resource at different points in time (*dynamic efficiency*), as is the case when deciding how quickly to extract a nonrenewable resource, such as fossil groundwater. The current level of resource use should be set so that the MNBs from future use of the resource are equal to current MNBs—except that future MNBs are discounted at a certain rate (the *discount rate*) to take into account the fact that benefits obtained now are worth more than benefits later. There is ongoing controversy over what discount rate to choose: too high a discount rate will lead to overextraction in the present, while too low a discount rate will lead to underextraction.

Economists argue that for many resources, free markets do a good job of achieving efficient allocation. Markets ensure that the resource is sold to the users with the highest total net benefits, since they are willing to pay the most for the resource. In addition, the market price of the resource—the reflection of the balance between supply and demand—will vary with changes in either supply or demand. Price thus serves as an important signal of the scarcity of the resource and regulates how much of it will be consumed. Markets tend to work best for *private goods*, defined as resources that are *rival*—my use of the resource precludes your use—and *excludable*—I can own the good and prevent you from accessing it (Table 12.1). Most goods that are traded in markets—from clothing to stocks to food—are in fact private goods.

However, economists also point out that there are important situations of *market failure*, in which markets do not give the socially efficient result. Market failures often result from some kind of *externality*, where the costs (or benefits) of an activity are not borne solely by the person carrying out the activity. For example, most of the costs of air or water pollution are borne by society as a whole, not just by those who generate the pollution. As a result, a firm deciding how many widgets to produce doesn't take into account all the pollution costs of widget production and is likely to overproduce relative to the socially optimal level.

Market failures are particularly common for resources—like water—that are not exclusively private goods. Some uses of water—such as navigation in large rivers or provision of in-stream habitat—are *pure public goods* (Table 12.1): nonrival (multiple users can share the river without affecting one another's share) and nonexcludable (it is difficult or impossible to control access). Other uses—such as many fisheries—qualify as *open-access resources*, that is, resources that are largely rival (my use precludes your use) but also nonexcludable. Depending on the legal doctrine being applied (see below), most off-stream water use functions as either a *common-property resource* (rival and partially excludable, since it is shared among a group of users who follow certain rules) or a *private good* (rival and excludable, since my right to use the water is well defined and absolute).

Open-access resources are the most likely to get overused, since each user gets all the benefits from his own use but shares the costs (e.g., depletion of the resource) with all other users. Hardin's famous "tragedy of the commons" (Hardin 1968) exemplifies how overgrazing of a common pasture (really an open-access pasture) can result from the decisions of individual users. Even the most economics-oriented observers would

TABLE 12.1

Economic categorization of resources, with water examples
for each category.

Note that rivalry is a natural feature of the resource, while excludability is a function of human
institutions that govern access to the resource.

| | | Rival? *(my use precludes your use)* | |
|---|---|---|---|
| | | Yes | No |
| **Excludable?** *(I can prevent you from accessing the resource)* | Yes | *pure private good* (water use in western US) | |
| | Partially* | *common property* (water use in eastern US) | |
| | No | *open access resource* (open-ocean fisheries) | *pure public good* (navigation, ecosystem health) |

* Partial excludability refers to a situation where a resource is shared (usually among a limited group of users) following a set of more or less well-defined rules or customs.

agree that we need adequate rules and institutions to prevent the tragedy of the commons in water use; the controversy comes in deciding what form these rules should take.

An additional problem with markets has to do with the fact that not all users of a resource necessarily have the ability to pay market prices. For water, for example, it seems self-evident that providing basic amounts to the poor (20–50 liters per person per day) provides higher social net benefits than providing that same water to rich households that are already using large quantities of water for nonessential uses. Yet the rich households may well be willing to pay more.

## •12.2. Laws and Markets for Allocating Water

Given a limited supply of water and many competing uses, how much water should each user receive? An ideal water allocation mechanism should accomplish four goals:

1. Clarity: Minimize uncertainty and the resulting conflicts between users.
2. Economic efficiency: Maximize the total net benefits of water use.

3. Environmental protection: Balance human and ecosystem use by accounting for the benefits provided by in-stream (environmental) flows[2] (see section 8.2).
4. Distributional equity: Distribute the benefits of water use in an equitable manner.

Implicit in these criteria is the goal of sustainability: when a renewable water resource is being allocated, use today should not interfere with future use; when a depletable resource (i.e., fossil groundwater) is being used, use today should be balanced with future use according to the principles of dynamic efficiency and intergenerational equity.

In this section, we look first at current rules governing water allocation and then examine how markets might—or might not—help us meet these four criteria. Our focus here is primarily on allocation among different users within a state or country. Here the competing users are typically different municipal water utilities, irrigation districts, farms, or industrial users that are drawing from the same surface water or groundwater source. There are two other kinds of allocation to consider: allocation among different countries, discussed in section 13.1; and allocation among different states within the US, discussed in section 13.2.

## Legal Doctrines for Water Allocation

There are two primary doctrines governing how surface water is allocated to different users: the riparian doctrine and the prior appropriation doctrine.

The *riparian doctrine*, which originated in the relatively water-abundant conditions of England and was subsequently adopted by the eastern US states, treats a river as common property, a resource to be shared among all the *riparians*, that is, those who own land bordering the river. Each riparian has an equal right to use water from the river (although in the strictest version of the riparian doctrine, this use must take place within the watershed). The original riparian doctrine prohibited users from significantly altering the "natural flow" of the stream. This worked well when demand was low relative to the size of the stream and when much of the use was nonconsumptive (e.g., watermills). However, as demands on the resource have increased, the riparian doctrine has generally abandoned the natural flow requirement and moved to requiring only

---

2. This is implicitly included in the previous criterion (economic efficiency) but is worth highlighting as a separate criterion because it is often neglected in practice.

that each use be "reasonable" relative to other uses—which does not really provide much guidance on how to share a limited supply.

The riparian doctrine fails to a large extent on all four criteria above. It can lead to conflict because of the lack of clarity on how to share shortfalls resulting from drought or growth in demand. It has only a weak mechanism ("reasonableness") for ensuring that more water goes to the users who will derive the greatest benefit from it. It has no mechanism for accounting for the benefits of water left in the stream; indeed, many streams governed by the riparian doctrine suffer from inadequate in-stream flows (Trout Unlimited 2006). The riparian doctrine is equitable in the limited sense of treating all riparians equally, but not in the sense of ensuring that nonriparians (e.g., residents of nearby cities) have enough water.

In more arid regions, such as the western US, the *prior appropriation doctrine* has dominated instead. This doctrine originates from the desire to encourage settlement of the West by providing land, mineral, and water rights to gold miners, farmers, ranchers, and entrepreneurs who could make use of these resources. It does this by assigning private property rights to the use of water (a *usufructuary* right)—though not to the water itself, which is still owned by the state, just as in the riparian doctrine. These private water rights take the form of an amount (or flow) of water, along with a *filing date*, which determines the level of seniority of the water right ("first in time, first in right"). When there is not enough water available to satisfy all the water claims, water rights get filled in order of seniority, with more senior users getting their full allocation before more junior users get any water. Water allocations must be put to "beneficial use" or they will be forfeited—the "use it or lose it" provision. The list of beneficial uses typically includes irrigation, domestic water supply, hydropower, and industrial use, but only recently is beginning to include in-stream uses for environmental flows.

The prior appropriation doctrine would seem to do a little better than the riparian doctrine on the clarity criterion, but in practice has not resulted in less conflict, probably because it applies in water-scarce regions, and because of numerous additional complexities that leave plenty of room for legal battles (see section 13.2). Like the riparian doctrine, prior appropriation has no mechanism for ensuring that water use is efficient or equitable and often doesn't take into account the benefits provided by in-stream flows.

Water law for groundwater is generally distinct from surface water law, despite the fact that the two are often hydrologically connected and should be managed conjunctively. Groundwater rights range from absolute ownership (*rule of capture*), in which there are no limits on how much groundwater a landowner can pump, to versions of prior appropriation, in which an annual pumping volume is specified for each user within an aquifer, to versions of riparianism (*correlative rights*), in which pumping rights are shared among the aquifer riparians based on criteria of reasonableness. The key to successful groundwater management—which is not necessarily included in any of these doctrines—is recognition of two basic hydrologic facts: (1) groundwater pumping on private property can affect both the water table in adjacent properties and the flow in nearby surface water; and (2) groundwater can easily be pumped at rates exceeding its recharge, but this is a depletable resource whose efficient utilization needs to take into account future users as well as current ones.

Recently, two opposing proposals have surfaced on how to move forward in improving water allocation: on the one hand, a move toward markets and on the other hand, a shift toward greater governmental control. Both reflect the fact that in our age of increasing water scarcity, we are desperate for new systems that are up to the task of distributing water efficiently and fairly. Both also reflect the search for ways to move water from low-valued but senior agricultural uses to newer urban and environmental uses. The next two sections discuss these approaches.

## Water Markets for Improved Allocation

Economists have repeatedly pointed out the inefficiency of current water allocation doctrines. For example, Wahl (1989) examined the many hidden subsidies in the Bureau of Reclamation's water pricing policies and found that if irrigators had to pay the full cost of water, many irrigation projects in the West would not have been viable; these projects should never have been built. Holland and Moore (2003) calculated that the Central Arizona Project (the canal transporting Colorado River water to Arizona cities and farms) causes a net economic loss of $2.6 billion; it, too, should not have been built.

Many economists believe that water markets can overcome the shortcomings of current allocation doctrines by allowing the "invisible hand" of the marketplace to figure out where the available water can best be used. For example, Booker et al. (2005) built a model of water use in the Rio Grande Basin and found that under drought conditions, water markets would reduce economic damages by 33% relative to current water allocation rules.

Several complications to the creation of water markets must be acknowledged:

- To create a market, one first needs clearly defined, transferrable individual water rights. Markets thus tend to be more compatible with the prior appropriation doctrine, which has at least the beginnings of individual water rights.
- To market water, it must be possible to move it from one user to another. The cost of doing this (often described as a type of *transaction cost*) is relatively high for water, since it is a bulky, heavy resource. For this reason, the markets that have developed for water have tended to involve users within a relatively small area.
- Water sales often involve impacts to third parties (externalities). As discussed in Chapter 3, a large part of any water use is nonconsumptive and ends up returning to the river to be used again downstream. If I sell my water right for use elsewhere, I will be shortchanging the downstream users who were dependent on my return flow. Even if those users are junior users under the prior appropriation doctrine, I am not allowed to harm their existing rights. This was confirmed in 1972 by the case of *City of Denver v. Fulton Irrigating Ditch Co.*, in which downstream users were able to prevent Denver and Coors Beer from executing a water trade. One way around this problem is to allow the transfer of only the fraction of the water right that went to consumptive use, although this is often hard to define.
- A related kind of third-party impact concerns the effects of water sales on the economy of the selling area—so-called area-of-origin issues. Even when the overall effect of a water sale is positive, its local effect may be negative. (Or, put another way, markets may be efficient but they are not necessarily equitable.) The political opposition to such transfers can be overwhelming.
- More broadly, the concept of "selling our water" has a very bad reputation. It connotes subterfuge and shortsightedness, and tends to evoke very strong negative reactions in at least some segments of the public, in part because of the history of water grabs, for example, in the US West.
- Markets will protect in-stream flows only to the extent that there is someone willing to pay for them. Since in-stream flows are a public good, one would expect that they would be undervalued by a market. Nonetheless, there are environmental groups that see

markets as their best chance to protect streamflow and have been actively buying water rights in order to convert them to in-stream flows. Scarborough and Lund (2007) even opine that "no approach offers as much promise to restoring stream flows as transfers of water through markets," although many would disagree.

Because of these barriers, some have argued that true water markets are unlikely to emerge and that so-called water markets are really "regulatory management masquerading as a market" (Dellapenna 2007). The California Water Bank is offered as an example: the state determined who could sell, who could buy, and what the price would be, essentially using the water bank as a way of accomplishing its predetermined goal of transferring water from agricultural to urban users.

Nevertheless, there are cases—particularly in Australia, Chile, and the US West—of water markets that exhibit greater flexibility and freedom, although still with tight government oversight. We briefly examine each of these now.

Australia's Murray-Darling basin has had water trading since the late 1980s, when emerging scarcity led to legal changes designed to encourage water markets. The development of formal market institutions in the late 1990s, such as the Northern Victoria Water Exchange (now known as Watermove), simplified the trading process and allowed an increase in trading volume (Bjornlund 2003). Trades of water generally are limited to agricultural users within the same geographic zone and are most commonly short-term sales of a given volume of water, rather than permanent transfers of water rights. Nonetheless, the market is quite active (in 2001–2002, 18% of water use in the Goulburn system was involved in the market; Bjornlund 2003) and has been shown to produce an increase in economic efficiency (Brooks and Harris 2008).

Chile has gone the farthest down the free-market path, at least in legal structure. Its 1981 Water Code established private rights in water (separate from land rights), allowed laissez-faire trading of these rights, and restricted government involvement in water management. In practice, however, Bauer's (2004) review of the evidence concludes that "examples of significant market activity . . . remain limited to a few areas of the desert north and the metropolitan area of Santiago" and that the Water Code has not resulted in significantly improved efficiency. This may indicate in part that water markets need greater government involvement—in establishing institutions, reducing transaction costs, maintaining water transfer infrastructure, and regulating externalities.

In the US West, the prior appropriation system allows for transfers of water, subject to government approval and in the absence of harm to other users. These transfers may be leases of varying durations (1–100 years) or permanent sales of water rights. Donohew's (2009) compilation of water transfers from 1987 to 2007 shows that there has been an increase in the number of transactions over time. Most of the 3675 transfers recorded have been permanent sales. Although sales tend to involve smaller *annual* volumes than do leases, they represent the majority of the *cumulative* volume. The biggest category of transfers is from agriculture to urban users, followed by agriculture to agriculture, urban to urban, and agriculture to environment. The total 21-year volume of water transferred is 187 km$^3$, or about half the total flow of the Colorado River over this time period.

## Regulated Riparianism for Improved Allocation

The other emerging approach to fixing the allocation system involves greater—rather than less—government involvement. This approach, which has been termed *regulated riparianism,* merges the clearly defined water right of the prior appropriation doctrine with the reasonableness of the riparian doctrine, while adding in a level of integrative government review. Specifically, users must apply for permits, which are issued by a government agency based on criteria of reasonableness and, ideally, a comprehensive review of water supply and demand in the basin. Permits are issued for a defined time period, after which they must be renewed; this gives the opportunity for periodic reevaluation, while still allowing secure water rights for a reasonably long time. Some version of this approach is becoming more popular, especially in the eastern US, and the American Society of Civil Engineers has adopted a model code along these lines.

In some sense, this approach returns to the *public trust doctrine* that underlies both riparianism and prior appropriation: the notion that natural waters are a resource that belongs to the public as a whole. The great promise of regulated riparianism lies in the possibility of comprehensive watershed planning, with the "visible hand" of government weighing different needs—including in-stream flows—and achieving an allocation that balances efficiency, equity, and environmental protection. This is in contrast to the piecemeal allocation that is typical of both riparian and appropriative rights, in which there is no opportunity for holistic review of individual decisions accumulating over time.

But is government up to this challenge? One's answer to this no doubt depends on one's general sense of what government can accomplish. There are certainly reasons to doubt whether the government agencies that brought us current water management—with its many examples of inefficiency and inequity—can really pull off the complex goals outlined above.

Here are some conditions that should be in place to allow government allocation to succeed:

- Clear mission and authority: A new or existing government agency needs to be assigned comprehensive water allocation as its primary mission and given the clear authority to carry out its task. This may prove difficult, given the large number of state agencies that typically have competing water-related agendas.
- Balance: Those making decisions on water allocation need to be able to consider and balance the needs of different constituents (water utilities, industry, environment, hydropower), without being beholden to any one group. They should be guided by clear rules on how to achieve that balance.
- Transparency and accountability: The decision-making process should be open to public examination and comments. Corruption can undermine the ability of government to carry out this task.
- Funding: The relevant government agency needs to be provided with sufficient funding and staffing to carry out its job.
- Data: In order to allocate water fairly, the agency must have good data both on natural flows and on how much water is being used by different entities. This requires comprehensive environmental monitoring and water use metering programs, and a good database for storing the results.

As we move forward with both markets and regulated riparianism in different places (and sometimes in the same place), it will be fascinating to see which approach brings greater success in achieving the goals outlined at the beginning of this section.

## 12.3. Pricing for Conservation, Cost Recovery, and Fairness

How much should end-users, such as households or farmers, pay for water? Is there a way to use pricing to achieve more efficient water use?

### Theory

Some people believe that water should be free. After all, it falls from the skies as a gift from the hydrological cycle and flows downhill to us without cost. If natural waters belong to the public, how can water utilities get away with charging us for what was ours to begin with?

Upon reflection, though, most people realize that what is being delivered from their taps is not the same as what falls from the sky. The water company must extract the water, treat it, and deliver it, all of which costs money. Moreover, the water company must construct and maintain an extensive system of infrastructure—pumping stations, treatment facilities, miles and miles of pipes, thousands of water meters—which also costs money.

To an economist, the ideal water price would include not just the costs of treatment, delivery, and maintenance, but also the value of the natural water itself. The fact that water falls from the sky doesn't make it worthless, just as the fact that oil occurs naturally doesn't mean that it is without value. The value of oil rises and falls in response to changes in demand and supply, with the market price increasing as oil becomes more scarce; shouldn't the same be true for water? The ideal price of water, then, would equal the *marginal cost* of water delivery (the cost of delivering the last increment of water), including both supply costs (the costs incurred by the utility, including maintenance costs) and the value of the raw water (often expressed as the *opportunity cost*—the value of the water for other possible uses).

Yet water is different from oil in several ways that make pricing it at its true cost very difficult:

- Water is essential for life. Applying coldhearted economic calculations to such a basic necessity seems wrong to many people. Yet economists would point out that food and shelter are also essential for life, and we nonetheless set prices for those goods in the marketplace. The counterargument, of course, is that food prices in many countries are kept artificially low through subsidies to farmers or through policies designed to keep staple foods affordable. Still, many countries deal with the equity issue not by lowering food prices but by providing direct assistance to those who can't afford to buy food. Shouldn't we do the same for water: have the market set prices and then provide assistance to those who need it?
- Water delivery is a natural monopoly. It doesn't make sense for multiple companies to lay parallel water pipes to serve the same area. Thus, water prices can't be set by competition among different suppliers. As is the case for other natural monopolies, government oversight of water rates is critical.

- Water has uses that are hard to value. The value of water to support in-stream habitat is undoubtedly greater than zero, but it is hard to put a number on.
- Water is publicly owned. For cultural, historical, and emotional reasons, there is a consensus in most societies that private ownership of water sources (as opposed to use rights) is wrong. If that is the case, then the value of raw water at its source (prior to treatment and delivery) should be captured by the public, not by water utilities. How do we charge customers the marginal cost of water—including the value of the raw water—without essentially giving away the raw water to the utility?

## The Argument for Higher Prices

Despite the difficulties of pricing water to achieve perfect economic efficiency, most economists would argue that modifications to water prices can help move us in the general direction of greater efficiency. In particular, current water prices are thought to be too low, at least in developed countries, leading to overuse. If utilities raised water prices to levels more closely approximating the true value of water, household water use would drop and water would be freed up from wasteful household uses and transferred to uses that provide greater societal value, like keeping streams from going dry.

How much water could be saved by price increases? The standard measure of the responsiveness of demand to price is the *demand elasticity*, defined as the percentage change in demand that results from a 1% change in price. Several studies have found that residential water demand in the US generally has an elasticity of about −0.4 in the short run (i.e., a 1% increase in price results in a 0.4% decrease in demand; Olmstead and Stavins 2009). Water demand elasticity tends to be higher at higher prices and over longer time periods, the latter because long-term responses include technological as well as behavioral change. Elasticity is also increased by water bills that specifically list marginal prices, thus making consumers more aware of the price that they are paying (Gaudin 2006).

How do price increases compare to other types of water conservation measures in terms of economic efficiency? Mansur and Olmstead (2007) used data from 11 North American cities to compare price and non-price conservation measures. They found that, compared to mandatory restrictions on outdoor watering, price measures can lead to the same level of reduction in water use but at a lower social cost. This is because watering restrictions impose uniform reductions on all consumers, while price increases result in differential responses, with the greatest reduction coming from consumers who value the water the least (and in the types of uses that are valued the least). Not surprisingly, the consumers who reduce their water use the most are the ones with relatively low incomes, which raises equity concerns—although these are mitigated somewhat by the fact that none of the uses being cut is essential.

From the perspective of water utilities, there is another goal for price increases that generally looms larger than the goal of efficiency: cost re-covery. Most water utilities rely on water tariffs from customers for the bulk of their revenue, although public utilities may also receive funding from the government. When water prices are too low, utilities may not be able to cover their costs and may have to cut back on their level of service or defer maintenance on essential infrastructure.

In some cases, especially in developing countries, underfunded utilities experience a vicious cycle of inefficient management: they don't have the capital to provide good service, install meters, or collect tariffs; as a result, they can't charge high rates, gain additional customers, or even collect from the customers that they do serve. In addition, they often lose a great deal of water to leaks and illegal connections in their poorly maintained infrastructure—water that they are not getting paid for.

In the US, in contrast, utilities typically are required to achieve cost recovery—but in ways that prevent them from sending appropriate price signals to consumers. Specifically, most water utilities in the US are re-quired to set rates based on their *average* cost rather than their marginal cost. When additional, expensive sources (like desalination) come online, they are averaged into the overall costs of the utility, resulting in a muted price signal to the consumer that doesn't accurately convey the high cost of this new source (NRC 2008).

In any case, raising prices can help water utilities with their bottom line. In fact, since elasticities are almost always smaller than 1 (in absolute value), raising prices results in greater revenues, because the increase in price more than makes up for the decrease in demand. This contrasts with other conservation measures (e.g., installation of low-flow fixtures), where utilities can end up in the awkward position of having to raise prices after a successful conservation campaign, because demand has dropped enough that the utility cannot meet its fixed costs.

## The Current Pricing Picture

What do current water prices look like, and how do they compare to the economically ideal prices? Unfortunately, good data on both questions are difficult to find—the former because of the great diversity among localities in pricing and the latter because of the economic complexities of evaluating true marginal costs. Yet some salient facts can be assembled:

- Residential water utilities use a variety of pricing structures: flat fees (a fixed charge independent of the amount of water consumed, usually used because of the absence of water meters), uniform volumetric rates (a set rate per unit of water consumed, often coupled with a fixed charge), decreasing block tariffs (a price per unit that decreases as the volume of water used increases), and increasing block tariffs. At least in countries that are members of the OECD (Organisation for Economic Co-operation and Development, a group of 30 industrialized nations), there appears to be a shift away from flat fees and decreasing block tariffs toward the more economically sound uniform volumetric rates and increasing block tariffs (OECD 2009).

- In OECD countries, water utility prices—including both water and wastewater—range from about $0.70/m$^3$ to more than $4/m$^3$ (OECD 2009). Prices charged by utilities in developing and transitional countries also span a huge range but are often somewhat lower (e.g., India: $0.10/m$^3$; Argentina: $0.20/m$^3$; China: $0.40/m$^3$; Philippines: $0.70/m$^3$; Indonesia: $0.80/m$^3$; Chile: $1.20/m$^3$; OECD 2009), in part because of the lower level of wastewater treatment.

- Water prices charged by utilities have generally been increasing (in real terms) throughout the world, as costs increase and as utilities come under greater pressure to recover their costs.

- Many people in poor countries—especially those living in urban slums—are not connected to a water utility and instead obtain their water from informal vendors, at rates that are typically 8 to 16 times as high as public utility rates (UNDP 2006). Thus, the highest rates are paid by those who are least able to afford it and who are receiving the lowest quality of service.

- Affordability is a serious issue. On the one hand, water and wastewater bills in OECD countries (OECD 2009) represent only 0.2–1.4% of average household income (0.3% in the US), which is generally much less than other utilities such as electricity or natural gas. On the other hand, the poorest 20% of the population in developing countries often spend 3–11% of their income on water (UNDP 2006). For the poor, safe water may be truly unaffordable, forcing them to use unsafe sources or to use less than they need. Even in OECD countries, the poorest 10% of the population may spend a significant fraction of their income on water (US: 2.2%; Germany: 3.5%; Poland: 9.0%; OECD 2009). An often cited target for affordability is <3% of household income.

- Economists believe that prices charged by utilities are still generally much lower than economically efficient prices. For example, Renzetti (1999) found that average residential water prices (not including wastewater) in Ontario were $0.32/m^3$, while marginal costs—not including raw water value—were $0.87/m^3$.

- Despite the many benefits that raising water prices would bring, there can be a great deal of public opposition to higher water prices, in both industrialized and developing countries. In my view, this opposition is largely unjustified in industrialized countries: water is a valuable commodity and people who can afford it should be willing to pay what it really costs. For the poorest communities, however, higher water prices pose a serious equity problem.

## The Equity Problem

How can we reconcile higher water prices with equity considerations? After all, nobody should be priced out of using water for the basic necessities of life (drinking, cooking, washing, sanitation). It is unconscionable that poor communities around the world are being asked to pay prices that they can't afford for something that is a basic human right. This situation is made even more disturbing by the fact that water is one of the main factors reinforcing the vicious cycle of poverty: people who can't afford safe water and sanitation are unable to get out of poverty, both because they suffer from frequent illness and because they must spend their time getting water rather than making a living.

The fundamental solution to this problem lies not just in the water realm but in alleviating poverty more broadly. In many cases, the real problem in poor communities is not that the water prices are too high but that the people are too poor.

Still, given the realities of global poverty, how should we deal with water pricing in poor communities? Three solutions seem promising.

First, providing access to an improved water source for those who lack it makes great sense from both an efficiency and an equity perspective. Even paying the full marginal costs of piped service is cheaper than paying a street vendor. Providing an improved water supply lowers both cost and price, improves quality of service, and has large additional economic benefits in avoided health-care costs and loss of productive time (see Chapter 9).

The second solution involves "lifeline" rates, where the amount of water necessary for health (perhaps 50 liters per person per day) is priced at a very low rate, while all water use above that level (the vast majority of water use in most households in industrialized countries) is priced at a much higher rate.

A third approach involves some form of rebate or "water stamps," similar to food stamps, which are given on the basis of financial need and can be used to cover basic water costs. Implementing any of these solutions requires strong government oversight capacity, which often does not exist in the poorest countries.

## 12.4. Privatization

Water and sewer utilities can be either privately or publicly owned. Many people are unaware of which type of utility serves their own household. Yet ownership structure can potentially affect the efficiency and equity of service provision, and a great deal of controversy has recently attached to the issue of increasing privatization.

### The Opposing Positions on Privatization

The relative importance of private and public water utilities varies by country (Figure 12.1), with private ownership being most dominant in Chile, France, and the UK. In many countries, the balance between public and private ownership has shifted over time. In the US, early-nineteenth-century water providers were almost all private, but there was a gradual shift toward public water provision as cities grew and the public interest in water and sanitation became more obvious. In the last several years, there has been a move back toward privatization in some cities in the US and around the world. By one estimate (Barlow 2007), the number of people worldwide who get their water from a private utility increased from 50 million in 1990 to 600 million in 2006.

As discussed in Chapter 9, the interest in involving the private sector in water and sanitation in developing countries stems partly from the large amounts of capital that are needed to meet the water MDGs, and the fact that funding from donor countries has fallen well short of these amounts. As can be seen in Figure 12.2, private investment in the water and sanitation sector has picked up some of the slack in aid.

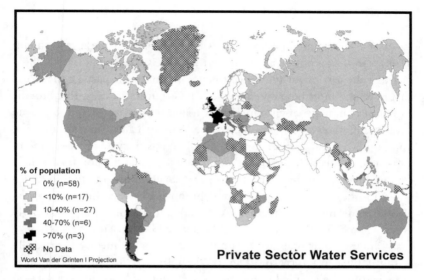

*Figure 12.1.* Map of private sector water service provision. The number of countries in each category is shown in the legend. Data source: Edouard Pérard, World Bank, personal communication. Map produced by Stacey Maples, Yale University.

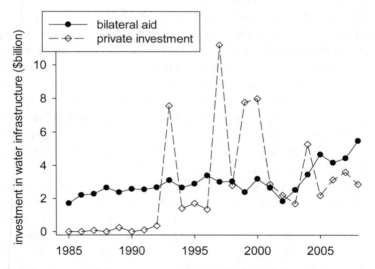

*Figure 12.2.* Investment in water supply and sanitation from bilateral assistance (e.g., US foreign aid) and private investment. Data on bilateral assistance from OECD (stats.oecd.org/qwids/); data on private investment from World Bank (ppi. worldbank.org/). All data are in 2007 dollars. Data are not readily available on two additional sources of funding: multilateral assistance (e.g., World Bank) and the recipient countries themselves (governments and/or users).

At the same time, the companies that are making these investments are in it to make money, not to meet the MDGs. This is illustrated by the fact that the vast majority of investments are not in the parts of the world that need the most help meeting the MDGs but in the areas that provide the most conducive environments for doing business: 88% of the private investments shown in Figure 12.2 have gone to Latin America and East and Southeast Asia (especially China), while less than 0.5% have gone to sub-Saharan Africa. It should be noted, however, that the companies most heavily involved in these investments are a handful of large multinational water utilities based in Europe, most prominently Suez Environnement, Veolia Environnement (formerly Vivendi), and Thames Water (formerly RWE Thames, now owned by the Macquarie Group). More recently, several Chinese companies are becoming more heavily involved, especially within China (Izaguirre and Pérard 2009). In any case, these companies are investing their capital in the hopes of recouping it—with a profit—in the form of user fees (and, in some cases, government payments). Thus, these investments should be seen ultimately as user funding, though with the private company providing the large sums of up-front investment needed to build new infrastructure.

Besides capital, the private companies also bring something else to the table: the technical capacity and expertise to build and operate large water systems. Given the inadequate technical and managerial capabilities of government in many developing countries (not to mention corruption problems), the professionalism of the private sector is often seen as a key factor in improving water and sanitation services. In fact, some argue that the efficiency advantages of the private sector are so large that these companies can generate a handsome profit for their shareholders while still providing better service at a lower cost than the public sector in developing countries can hope to offer.

Nonetheless, there has been very significant opposition to the wave of privatization. In large part, this opposition stems from an anti-globalization perspective, which sees Suez and the other European multinational water corporations as unfairly reaping financial benefits by exploiting unregulated markets in poor countries without regard for equity or sustainability. This charge has particular force for water, given several factors: water's status as a basic human right, water's sacred role in many cultures, the perception of water as a free good owned by the public as a whole, and the ease with which water sources can be permanently damaged by depletion or pollution. Can we count on the short-term thinking of the private sector to protect our vital water sources for the long-term good of the public? Adding to the tension is the fact that in many cases the World Bank or the IMF has essentially forced countries to accept water privatization as part of "structural adjustment" policies that have in themselves generated considerable opposition.

In fact, anti-privatization activists see a conspiracy of corporate interests, northern governments, and international groups (World Bank, UN, etc.), all combining forces to push privatization down the throats of poor countries, even as rich countries maintain their tradition of largely public ownership. In this view, the goal of the "water cartel" is to seize control of increasingly scarce and valuable water supplies. In the process, the argument goes, they exacerbate, rather than solve, the crisis in water and sanitation, by diverting aid money to privatization projects and preventing local governments from implementing sustainable local solutions.

Battles over privatization are not limited to the developing world, though the battles there have been the hardest-fought and best-publicized ones (Box 12.1). Several American cities—including Atlanta, New Orleans, and Stockton, California—have seen privatization attempts that failed at some stage in the process.

## Moving Forward: A Middle Ground

The reality of privatization is more complex than the simple picture of rapacious corporations out to make a quick profit. But privatization is not a panacea that will solve all our water problems.

There is no simple answer to the basic question, Which is better, private or public ownership? In theory, each structure has certain advantages. For example, public entities, especially in industrialized countries, can borrow money more cheaply, can raise money through taxes, and are more directly accountable to the public. On the other hand, as discussed above, private companies can bring a large capital base, technical expertise, and the ability to creatively minimize costs. In practice, studies have generally not found systematic differences in overall efficiency between the two structures (Pérard 2009).

In addition, it is important to realize that the sharp distinction between private and public is an artificial construct. In reality, there is a spectrum of ways that private actors can be involved in water service provision. These include the following, from least to greatest involvement:

- Private purchases of bonds sold by public utilities
- Management contracts, in which a private company is paid a fee for operating a publicly owned utility

## Box 12.1. Cochabamba and Its Lessons

The wave of opposition to privatization found perhaps its most dramatic expression in the "water wars" in Cochabamba, Bolivia, in early 2000, in which a 17-year-old demonstrator was killed by the military and the Aguas del Tunari consortium (spearheaded by Bechtel), which had been hired to run the city's water supply, was ultimately forced to withdraw from its contract. This was widely viewed as a stunning victory for a people's movement that had successfully fought off a large multinational, their own government, and the World Bank. Yet Cochabamba illustrates some of the complexities of the privatization issue:

- *Population growth, poor infrastructure, and scarcity*: Aguas del Tunari took over a water and sewer system that was completely inadequate to deal with the rapidly growing population of Cochabamba. Much of the city did not have access to water and sanitation, and where pipes did exist, they were in poor condition, losing almost half of their flow to leaks. In addition, the city had nearly exhausted its nearby water sources.
- *Corruption*: One of the conditions of the contract between Bolivia and Aguas was a requirement to bring additional water to the city by completing the Misicuni Dam project—not because this project made economic sense, but because of pressure from Cochabamba's mayor, who relished the opportunities for political patronage that the expensive project offered.
- *Rising water bills*: Aguas del Tunari immediately raised water bills dramatically in order to start covering the costs of investments in infrastructure. Although there is some dispute over exactly how large this increase was, two facts seem clear: the additional costs were a significant burden to poor households (in some cases representing a quarter of their monthly income), and bills rose *before* people saw any significant improvements in service.
- *Poorly structured contract*: Aguas del Tunari's contract gave it complete ownership of the water sources and did not provide for effective government oversight.
- *No attention to local communities*: Before the privatization, several neighborhoods in Cochabamba were served by well-functioning small-scale water cooperatives, which had installed groundwater wells and built neighborhood distribution networks. This infrastructure was given over to Aguas del Tunari without any consultation with local communities.

- *The difficulties of public management:* Since the withdrawal of Aguas, Cochabamba's water and sanitation services have, by most accounts, stagnated under public management, with much of the city (especially poorer residents) still without a reliable water supply. The people's victory may prove to be a hollow one.

- Leases, in which a private company pays to lease the existing assets of the utility for a set period, during which it operates and maintains the system and receives revenues
- Concessions, in which the private company—in addition to operating the system—makes capital investments in new infrastructure but does not acquire long-term ownership of any assets (also known as build-operate-transfer contracts)
- Divestitures, in which the utility assets are transferred to a private company, sometimes for a fixed period of time

An additional step, in which ownership of the water source itself is transferred to a private company, should be considered unacceptable. It is also important to realize that the farther one moves along the privatization spectrum, the more crucial it is that government play a strong regulatory role in setting rates, ensuring sustainability, controlling water quality, and so on. Thus, water provision by its nature requires both public and private involvement—*a public-private partnership.*

How do we determine which mix of private and public involvement is suited for a particular situation? Peter Gleick has suggested a set of principles for successful and equitable privatization (Table 12.2). The fundamental problem, however, as he and others have pointed out, is that the locales that are in greatest need of privatization—those with weak governance and technical capacity—are precisely those where privatization is most likely to go wrong, since government doesn't have the ability to implement these principles. In the absence of strong oversight, counting on the private sector to behave well on its own is not generally a good idea. Besides pointing to the need for strengthening governance, this phenomenon also suggests a potential role for the World Bank and other lenders in helping countries write and implement better privatization contracts. Also needed is some mechanism for getting the benefits of private involvement to the countries that are in greatest need of help but are least attractive to private companies, in particular many of the countries of sub-Saharan Africa.

TABLE 12.2
The Pacific Institute Privatization Principles
(from Palaniappan et al. 2004)

1. Manage water as a social good.
   1.1. Meet basic human needs for water.
   1.2. Meet basic ecosystem needs for water.
   1.3. Subsidize basic water use for the poor.
2. Use sound economics in water management.
   2.1. Rates should be fair and reasonable.
   2.2. Link rate increases with service improvements.
   2.3. Subsidies, if necessary, should be economically and socially sound.
   2.4. Require economic analysis of "soft" alternatives before investing in new supply projects.
3. Maintain strong government regulation and public oversight.
   3.1. Retain public ownership of water sources.
   3.2. Governments should define and enforce water quality.
   3.3. Spell out public and private responsibilities in contracts.
   3.4. Develop clear dispute resolution procedures.
   3.5. Require independent technical assistance and contract review.
   3.6. Keep decision making open and transparent.

## 12.5. Cost-Benefit Analysis

The final area of dispute we discuss in this chapter has to do with the use of economic tools to decide which water development projects make sense.

As previously noted, large infrastructure projects—dams, irrigation canals, water and sewage pipelines, river channel "improvements" and the like—were a big part of water management during the twentieth century. These projects were very expensive but also promised huge benefits—in water storage, water delivery, electricity production, and improved navigation. For many years, the decisions to proceed with these projects did not involve specific accounting of benefits or costs, but rather were based on the desire to utilize newfound technological capabilities, along with a sense that "progress" demanded the increasing control of nature.

Over time, however, many countries, including the US, began to require formalized economic *cost-benefit analysis* (CBA; also referred to as benefit-cost analysis) of a project before deciding whether to proceed. Such analysis involves quantification, in dollar terms, of all the quantifiable costs and benefits, followed by calculation of a summary metric, such as the *benefit/cost ratio*, or the *net present value* (NPV) of the project (the benefits minus the costs, calculated over time and discounted to the present). Such an analysis presumably will reveal which projects are part of an economically efficient package of water development. Specifically, projects with positive NPVs or with benefit/cost ratios greater than 1 will produce net economic benefits. Different alternatives (including the status quo) can be compared and the economically optimal course selected.

Three main shortcomings of CBA can be identified: equity issues, unquantifiable factors, and pork barrel politics. Economists tend to acknowledge these shortcomings but insist that CBA is still a helpful tool, while environmental and social activists tend to think that these shortcomings are fatal and that CBA has only a small role to play in improving water management decisions.

## Equity

A project with positive net benefits can still have losers—people who are worse off than they were before. While it may make sense to move ahead with such projects, it is important to be aware of the equity issues involved. Ideally, a CBA can be helpful in identifying which groups are net losers, and it should be possible to compensate those groups. In practice, this is not always the case. For example, as noted in section 7.1, the people most directly harmed by the dams—those who are displaced by the dam itself or the resulting reservoir—often do not benefit from the electricity or water supply that the dam offers and are often not adequately compensated for their losses. Because of the recognition that benefits to society as a whole may override costs to individuals, the WCD recommendations do not require the consent of those who will be displaced by a dam project, except when they are part of an indigenous tribe.

## Unquantifiable Factors

Some costs and benefits of a project may be very difficult or impossible to quantify. How large is the cost when we submerge a river rapid that is sacred to a native tribe? What is the value of an endangered species of fish that may go extinct if a river is dammed? What is the dollar value of in-stream flows that perform a variety of ecological functions, from sediment movement to water quality improvement? How well can we really predict even the factors that we think of as quantifiable, such as the recreational value of a river or reservoir?

CBAs have to make myriad assumptions about the values of things. These assumptions are often invisible to laypeople, who are simply presented with the bottom line and do not have the resources to investigate the fine print.

Most economists acknowledge that some costs and benefits are hard to quantify, and they propose several solutions. First, methods are being developed to more accurately capture *nonuse values*; for example, a survey method referred to as *contingent valuation* can reveal how much people are willing to pay to preserve rivers that they will never visit. Second, CBA can require compilation of unquantifiable costs and benefits alongside quantifiable ones. In practice, however, many CBAs do not pay enough attention to unquantifiable factors, and these factors are often effectively given a value of zero.

## Pork Barrel Politics

Large projects tend to attract corruption and political meddling, which can result in inaccurate and misleading CBAs. Of course, these factors have always played a role in deciding which projects get built, including many of the wasteful large infrastructure projects of the twentieth century that were not subject to CBAs. In fact, the CBA process was designed to fight against political influence by increasing the transparency and objectivity of decision making. However, in reality, CBAs are complex documents with many hidden assumptions. Without a great deal of work, it can be hard to determine whether these assumptions and calculations are truly objective or are selected with the desired final outcome in mind. In fact, the supposed objectivity and scientific nature of CBAs may, ironically, mean that they don't get the full scrutiny that they deserve. The Tellico Dam provides a good example of a misleading and inaccurate CBA (Box. 12.2).

Political scientists have identified one common pattern of improper influence, known as the *iron triangle*. It is named for its three main players: the legislative branch, in particular the relevant congressional committees; the executive branch, in particular the relevant government agencies; and local/national interest groups who will benefit from a project. All three have an interest in moving forward with projects, since these projects will typically provide them with increased power or wealth or both. Each of the three has leverage for influencing the other two groups: interest groups can get out the vote, provide campaign contributions, and help increase agency budgets; government agencies can green-light certain projects and as a result bring money to interest groups and congressional districts; members of congressional committees control funding for agencies and projects and can wield (or not wield) their oversight responsibility in ways that favor certain projects.

## Box 12.2. The Tellico Dam and the Manipulation of Cost-Benefit Analysis

A famous example that illustrates the ways that CBAs can be manipulated is the construction of the Tellico Dam by the Tennessee Valley Authority (TVA). In the late 1960s, the TVA was a dam-building agency that had run out of good places to build dams. It proposed to move forward with the Tellico Dam on the Lower Tennessee River and produced a CBA that showed a benefit:cost ratio of 1.7:1. However, it is clear that the CBA was fundamentally flawed (Plater 2004). In particular, annual benefit estimates of $1.44 million for recreation and $714,000 for shoreline development were absurdly high, given the large number of other reservoirs in the area. If anything, recreation should have been considered a net cost: the recreational value of the last remaining reach of free-flowing river was higher than the recreational value of one more reservoir. In addition, the cost side of the CBA included only the financial costs of the dam structure, leaving out (besides the lost recreational opportunities) the loss of fertile farmland, archaeological sites, and fish habitat.

Ultimately, the fight over the Tellico Dam came to center on the Endangered Species Act (ESA) and the endangered snail darter that was found in the river. As part of this process, the absurdity of the CBA was highlighted by one of the members of the God Squad,[a] who exclaimed that the Tellico Dam was "a project that is 95% complete, and if one takes just the cost of finishing it against the [total] benefits, and does it properly, it still doesn't pay!" (Plater 2004). Ironically, the Tellico case confirmed both the power of the ESA and the power of the iron triangle: the former in the Supreme Court ruling that the dam could not be finished since it would endanger the snail darter's habitat, and the latter in the late-night congressional rider that ultimately allowed the dam to be completed.

Could a CBA like this be conducted today? Probably (hopefully!) not in the US, because of much stricter government guidelines for performing CBAs and greater scrutiny from the public. But for projects in developing countries, where there is sometimes less scrutiny, CBAs may be misleading.

[a] The God Squad is a committee, created by the 1978 ESA amendments, that can override the ESA's protections in cases where the public interest in doing so is overwhelming. In this case, the God Squad voted unanimously to reject the dam.

When the iron triangle gets going, it can be very hard to fight its power, especially since its workings are typically hidden from view. The iron triangle can be especially successful when the losers from a certain policy or project are widely dispersed or poorly organized. Special interests can use the iron triangle to win out over the common interest, because the loss to any member of the public is small and remote, while the gain to the interest group is large and immediate.

## 12.6. Conclusion

This chapter examined four topics related to the role of economics in water management: markets, pricing, privatization, and cost-benefit analysis. All four are among the most polarized issues in the entire water arena.

I believe that the "economic approach" discussed in this chapter—water markets, higher prices, involvement of private expertise, and economic analysis of proposed projects—can bring real improvements to water management, improvements that we desperately need if we are to avoid scarcity and bring water to all. Yet there must be very strong safeguards in place to ensure that the benefits of this approach are shared fairly and that attention is paid to the needs of the most vulnerable. In order to succeed in this task, both national governments and international agencies must improve their governance and oversight capacities, as well as their ability to communicate effectively with local communities.

You will have to make up your own mind on these issues. My hope is that you will take from this chapter some sense of the urgency of the questions and the complexity of the debate.

The next chapter continues to explore the question of water allocation, particularly the potential for both conflict and cooperation when different groups need to share a water resource.

# 13

## Conflict and Cooperation: Transboundary and Intersectoral Water Management

As a scarce and vital resource, water can be the focal point of serious conflict between different countries, states, or users. Several recent headlines offer examples: Egypt threatens its upstream neighbors with military action should they take more water from the Nile. Farmers in Texas rage at Mexico for not supplying the water it is required to release into the Rio Grande. Malaysia and Singapore wage a war of words over the price of water sold by Malaysia to its water-scarce neighbor. Iraqis blame Turkey and Syria for reductions in the flow of the Tigris River. The Indian states of Karnataka and Tamil Nadu fight over the waters of the Cauvery River. Georgia, Alabama, and Florida go to court over their shared rivers. Farmers in California's Central Valley conduct a protest march against federal allocations of water for endangered fish.

Yet water can also bring communities and countries together in search of better, more cooperative management: Israel and Jordan sign a peace treaty resolving their water dispute and providing for storage of Jordanian water in an Israeli reservoir. Brazil and Paraguay put aside their historic disputes and jointly build a large hydroelectric project. California, Arizona, and Nevada reach a new agreement on how to share shortages on the Colorado River in order to make best use of the available water. The countries of the Senegal River basin jointly manage two dams and share the benefits of the river.

This chapter addresses the issue of conflict and cooperation both internationally and within the US. As part of this topic, we discuss the current legal structure for water allocation at different scales—an issue that we have already started to address at the intrastate scale in section 12.2.

## 13.1. International Water Conflict and Cooperation

How do countries interact over shared water resources? Can we identify ways to avoid conflict and allocate water between countries efficiently and equitably?

There are certainly many situations where different countries must work out how to share a water resource. A 1999 survey (Wolf et al. 1999) counted 263 international river basins,[1] which collectively account for 45% of Earth's land surface, 40% of its population, and 60% of its river flow.[2] These include both boundary rivers (rivers that form a border) and boundary-crossing rivers (rivers that flow from an upstream country into a downstream one).

Much less is known about the status of international groundwater, due to the greater difficulty of mapping the extent and flow patterns of aquifers. Eckstein and Eckstein (2005) argue that international discussion of shared groundwater has been hampered by policymakers' limited understanding of hydrogeology. They identify six types of shared groundwater, ranging from a fossil aquifer that underlies multiple states (Model F) to paired aquifers that discharge to a boundary river from its two sides (Model A). Each case provides different physical, political, and legal constraints.

### International Law

What guidance does international law give us about how countries should share these transboundary water resources? Although a body of customary international law on this subject does exist, the rules for water allocation between different riparian nations are considerably less well developed than the intrastate water allocation doctrines discussed in section 12.2.

Upstream countries on international rivers typically invoke some version of the *Harmon doctrine of absolute sovereignty*, in which a country has complete right to all flow generated within its borders. Downstream

---

1. International basins are defined as river systems where the main stem or a perennial tributary is shared between at least two countries.

2. These numbers are higher than for a similar study in 1978 because of the fragmentation of the Soviet Union, Yugoslavia, and Czechoslovakia.

countries usually invoke the doctrine of *absolute river integrity*, which states that a downstream riparian has the right to receive the natural flow into the country from upstream, undiminished in quantity or quality. Clearly, these doctrines are mutually exclusive and can lead to conflict.

The UN Convention on the Law of the Non-Navigational Uses of International Watercourses, approved by the General Assembly in 1997, tries to balance these competing doctrines by articulating principles to govern international water allocation. These principles largely derive from the Helsinki Rules of the Uses of the Waters of International Rivers, which were put forward by the International Law Association in 1966. While these principles are fairly well accepted as the basis for customary international water law, the Convention has not entered into force, since only 17 countries (out of the required 35) have ratified it.

The most important principle articulated by the Convention is the requirement of "equitable and reasonable utilization and participation" (Article 5), which states that each riparian has a right to use the watercourse, but not an unlimited right: this right is constrained by a "duty to cooperate" in protecting and developing the watercourse. Article 7 makes explicit the point that the duty to cooperate includes the obligation to "take all appropriate measures to prevent . . . significant harm" to other users, though it notes that sometimes such harm may be unavoidable, in which case mitigation and compensation are appropriate.

How, then, to balance competing uses? The Convention notes that "no use of an international watercourse enjoys inherent priority over other uses," though "special regard [should be] given to the requirements of vital human needs" (Article 10). Beyond this, the Convention lists seven factors that should be considered in balancing different uses (Article 6): biophysical constraints, socioeconomic needs, population sizes, effects of each use on other uses, current and potential uses, the economic and environmental impact of each use, and the availability of alternatives to a given use. Additional important principles articulated by the Convention are the "regular exchange of data" and the requirement for timely "notification concerning planned measures with adverse effects."

The Convention has been criticized for being long on principles but short on both specific standards and procedures for implementation (Albert 2000). It is best thought of as an articulation of general principles, whose specific application must be negotiated between riparians in a particular river basin. Even if it were to enter into force, the Convention by itself would be unlikely to provide the basis for legal action to condemn anything but the most egregious cases of unilateralism—and even then, it is hard to imagine any substantive enforcement action. Given the status of international law in general, an articulation of principles is perhaps the most that can be hoped for.

In 2004, the International Law Association approved an updated summary of international water law, known as the Berlin Rules on Water Resources. The Berlin Rules do not deviate in major ways from the principles outlined above. However, they expand their scope to include subnational waters (to the extent that international law applies to them) and place more emphasis on public participation, integrated management, and environmental sustainability.

More specific guidance for water sharing is sometimes provided by regional or river basin agreements, where they exist. Notable examples of regional agreements include the 1992 UN Economic Commission for Europe (UNECE) Convention on the Protection and Use of Transboundary Watercourses and International Lakes, as well as the terms of reference for the Water Sector Coordinating Unit of the Southern African Development Community. In addition, some 400 bilateral or multilateral water-related treaties or agreements have been signed for specific basins since 1820 (Wolf 2002), with 106 of the world's 263 international river basins having at least one treaty. These agreements vary widely in their strength, but the majority fall short of providing substantive guidance on water sharing and cooperative management (Wolf 2002).

## Water Wars or Water Treaties?

> The wars of this century have been on oil, and the wars of the next century will be on water . . . unless we change the way we manage water.
>
> —*Ismail Serageldin, World Bank, 1995 speech in Stockholm,*
> *quoted in Serageldin (2009)*

This statement is often quoted in the media[3] to make the point that increasing water scarcity can lead to international conflict over shared water resources. But is this an accurate portrayal of international water relations or, as it has been called, hydropolitics?

Two schools of thought exist on the ways that water affects international relations. The first, exemplified by the quote above, emphasizes the vital role of water in national security. If countries are willing to use violence to protect other natural resources, such as oil, we would certainly

---

3. Serageldin (2009) points out that his statement is often quoted incompletely, without including the critical "unless" qualifier.

expect them to fight over limited and dwindling supplies of water, given how critical water is in fueling all aspects of a nation's economy and standard of living. This approach draws on the *neo-Malthusian* perspective, which believes that population growth will inevitably lead to resource scarcity and conflict.

The second school of thought argues that water provides ample opportunities and incentives for cooperation as well as conflict. The very nature of water—its vital importance, its fluid nature, the ease with which it is polluted, the advantages of managing it on a watershed basis—means that the benefits of cooperation tend to ultimately win out. Simply put, water is too important to fight over. This approach is part of a tradition that believes that resource scarcity can be solved through a combination of cooperative management, technological development, and substitution of different resources.

Examples can certainly be brought to support both perspectives. Tensions over water scarcity and unilateral development plans have contributed to violent conflict in the Middle East, including arguably playing some role in leading to the Six-Day War between Israel and its neighbors. In contrast, the Indus Waters Treaty has allowed Pakistan and India to share their joint water resources relatively peaceably, despite the other conflicts between the two countries.

To test the two theories more systematically, attempts have been made to quantitatively assess the prevalence of conflict over water. The most widely known effort is that of Aaron Wolf and co-workers at Oregon State University. They constructed a Water Event Database (Yoffe et al. 2003), consisting of all available international interactions (1948 and on) over shared water resources, and ranked each event on a scale from $-7$ (highly conflictual; formal declaration of war) to $+7$ (highly cooperative; voluntary unification into one nation). They found that the vast majority (70%) of events were cooperative (Figure 13.1), and there were very few events receiving rankings more conflictual than $-2$ ("strong/official verbal hostility"). Of the 37 events with rankings of $-5$ and $-6$, 30 involved Israel and its neighbors (no events received a score of $-7$).

An alternative analysis of water conflict was carried out by Gleditsch and co-workers at the Peace Research Institute in Norway. Their analysis looked at violent conflict between nations (not necessarily over water) and showed that countries that share a river basin have a higher likelihood of conflict than other similarly situated countries that don't share water resources. Gleditsch et al. (2006) argue that tensions over shared water resources may express themselves through conflicts that don't specifically

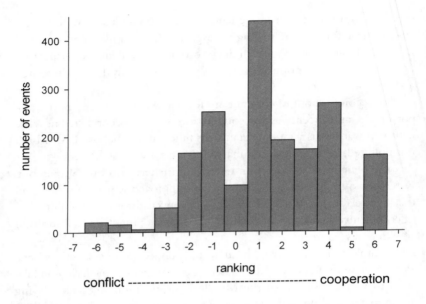

*Figure 13.1.* Number of international water events by conflict/cooperation ranking, from highly conflictual (–7) to highly cooperative (7). Data source: Wolf 2007.

involve water—conflicts that wouldn't be captured by the Wolf approach.

Zeitoun and Mirumachi (2008) also critique the Wolf approach, noting that conflict and cooperation are not necessarily mutually exclusive. Instead they argue for plotting events on a two-dimensional matrix that expresses the intensity of both conflict and cooperation for each event.

To my mind, the most important aspect of this debate is not whether water leads to conflict or cooperation—it can clearly lead to both—but *what factors* determine how nations interact over water and *how to best move* those interactions toward increased cooperation and better management.

## Factors Affecting Water Interactions

The literature on the factors determining water conflict and cooperation is full of contradictory results and diverging methodologies. The best we can do here is briefly discuss some of the factors that have been suggested to be important:

- Scarcity: Do more water-scarce basins have more water conflict? On the one hand, Giordano et al. (2005) argue that conflict over natural resources is most likely at intermediate levels of resource availability, since at low levels, there is nothing to fight over, while at high levels, there is no need to fight. On the other hand, Dinar (2009) argues that *cooperation* is most likely at intermediate levels of water scarcity: when water is abundant, there is no need for cooperation, and when it is scarce, sharing it in a cooperative way becomes difficult. Neither argument has strong support in data.[4] Another possibility is that scarcity may lead to both greater conflict and greater cooperation: as the value of water increases, it becomes important enough to be worth fighting over—but also important enough that the advantages of joint management become more salient. There is a hint of this logic in Yoffe et al. (2004)'s analysis, which shows that the most conflictive *and* most cooperative events both happen in arid regions (though this is contradicted by their finding that low annual precipitation tends to lead to conflict, while high annual precipitation tends to lead to cooperation).

- Political structure: Gleditsch et al. (2006), who examined only conflicts that led to at least one fatality, found that various political parameters were the most important factors predicting the likelihood of conflict over water. Conflict was much more likely for country pairs that had no history of peace, had unstable regimes, had relatively close capital cities, and had larger populations. Yoffe et al. (2003) also identified some of the same factors (high population density, hostile relations) as important.
- Internal politics: International conflict can be driven by internal power struggles, in which one side gains politically from maintaining an aggressive position regarding a transboundary water resource (Wolf 2007).
- Level of development: Yoffe et al. (2003) and Gleditsch et al. (2006) both found that conflict was less likely for countries with high per capita GDPs, with Gleditsch et al. (2006) suggesting that conflict may actually be highest at intermediate levels of development, when countries have the capacity to engage in conflict but not to solve their water problems technologically (e.g., with desalination).
- Rapid change: Wolf (2007) argues that an important factor leading to water conflict is rapid change—change in environmental conditions, socioeconomic circumstances, or political structures.

---

4. Dinar's statistical analysis is only significant for some of the outcomes analyzed, and in any case, the scarcity variable used (the Water Poverty Index) is poorly chosen, since it includes several other factors besides water scarcity.

- Institutions: Joint management institutions tend to increase the resilience of a river basin, that is, its capacity to absorb change without the creation of conflict (Wolf 2007).

## Tools for Moving Toward Cooperation

How do we move water relationships away from conflict and toward cooperation? Five key tools can be identified.

From rights to needs. Conflict tends to arise when countries focus on their contradictory claims of *rights* to use water from a joint resource. Shifting the focus toward the legitimate water-related *needs* of each country can lessen this conflict. This approach draws attention to the common humanity of all parties and allows creative thinking about ways to meet those needs with less water. Unless implemented carefully, however, allocations based on needs can perpetuate existing gaps in levels of development, since more developed economies can claim the need for more water to support irrigation and industry. Recognizing the basic human right to water for health and development must be part of a needs-based allocation.

Systems approach. One of the keys to water cooperation is the recognition that water sharing is not a zero-sum game. There are opportunities in cooperative water management to increase the size of the water pie, by achieving efficiencies that are not possible when a river basin is managed in a fragmented manner. One simple example involves the Nile River, where massive quantities of water are lost to evaporation from Egypt's Aswan reservoir. True basin management could have saved much of this water by locating the reservoir in the highlands of Ethiopia instead of in the desert of Egypt. Yet this proposal is unthinkable in the context of current nation-state water management. It should be noted that the benefits of the systems approach must be shared among all the riparians in order to provide incentives for moving in this direction.

The replacement value of water. Part of the power behind water conflicts is the notion that water is different from other scarce resources in that it has no substitutes. But in reality, there are substitutes for many water uses—ways to achieve the same result with less water. In addition, desalination can produce unlimited amounts of water, though at high cost. Thus, ultimately, nations fight over water because it is valuable, not because it is irreplaceable. In many cases, in fact, the economic value of the water at stake is less than the costs of conflict, whether those costs are the lost opportunities for systems management or the costs of military preparedness. Economic analyses can put these costs in perspective. For example, Fisher et al. (2005) point out that the cost of desalination sets the maximum value of the water resources that are in dispute between Israel and its neighbors. The disputed resource can then be seen to be worth a relatively paltry sum, compared to the myriad costs—economic, social, and environmental—of the ongoing conflict.

Virtual water. As discussed in section 7.4, trade in virtual water in the form of agricultural products can help alleviate regional scarcity and thus mitigate conflict. Conflict over water in the Middle East would be even more serious than it is today if this region were not an importer of virtual water. Conversely, one solution to the continuing water conflicts in the region may lie in a greater reliance on agricultural imports, with the water saved from the agricultural sector being used to ensure a more equitable and sustainable distribution of water among the parties. Ironically, the continuing conflicts in the region make countries less likely to give up their goals of food self-sufficiency.

Agreements and institutions. Water agreements need to have the flexibility to adapt to changing conditions. Key elements of a successful agreement include allocations that explicitly outline how shortages (e.g., due to drought or changing climate) will be shared, powerful joint institutions in which professionals can work together, mechanisms for dispute resolution, and data sharing requirements.

## Game Theory

Additional insight into the factors at play in international water disputes has come from the field of game theory. Several games, including those of Prisoner's Dilemma and Chicken, have been used to model a country's decision on how to act regarding transboundary water.

In the classic two-player game of Prisoner's Dilemma, each player is confronted with a choice of whether to "cooperate" with the other player or "defect." The payoff matrix governing the game is shown in Table 13.1. Key to this game is the fact that no matter what the other player does, you will do better by defecting. If your opponent cooperates, you get 4 points by defecting and 3 by cooperating; if your opponent defects, you get 2 by defecting and 1 by cooperating. This realization tends to drive both players to defect, resulting in payoffs of 2 to both players (bottom right corner). Indeed, this is the only *Nash equilibrium* in this game; that is, it is the only outcome where no player can gain by unilaterally changing her strategy. But is this outcome the desirable one? Clearly not. Both players would do better if they were to both cooperate (upper left corner, payoffs of 3 to each player instead of 2).

We should probably define "desirable outcome" a bit more precisely. One way to do this is to talk about *Pareto-optimal* outcomes, defined as those where we can't improve the result for one player without making it worse for the other player. By that criterion, any of the three outcomes *other* than the bottom right are Pareto-optimal. That is, in the game of Prisoner's Dilemma, the Nash equilibrium outcome is precisely the one outcome that is not optimal, in the sense that both players can do better! However, in order for both players to do better, they have to agree to cooperate; if only one cooperates and the other defects, the cooperator is worse off.

The Prisoner's Dilemma thus captures the *collective action problem*: how

TABLE 13.1
Payoff matrix for Prisoner's Dilemma

Each player chooses whether to cooperate or defect, with player A's choices represented by the columns and player B's choices by the rows. Each of the four cells of the table then represents an outcome of the game, with the payoff to player A indicated by the number above the diagonal line and the payoff to player B indicated by the number below the diagonal line. Higher payoffs are better.

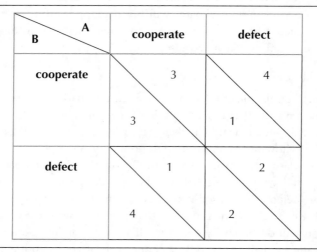

do we get both players to cooperate for a good outcome, rather than defect out of either fear or greed? (Fear motivates defection by raising the specter of the other player defecting and leaving you with a payoff of 1 unless you defect. Greed motivates defection by encouraging the hope that the other player will cooperate and you will get a payoff of 4 from defecting.)

From my experience playing this game with students in my Water Resources class, I can attest, as others have noted, that two modifications to the game tend to move the outcome of the Prisoner's Dilemma toward dual-cooperation: playing an *open game* and playing a *repeated game*. In an open game, where each player can see what the other player is doing and adjust accordingly, both players tend to realize that their best outcome will result from cooperating. Similarly, when the game is played repeatedly with the same partners, cooperation tends to dominate, since defection by one player can be punished by defection by the other player on the next turn (a tit-for-tat strategy).

What does this have to teach us about transboundary water management? One can construct scenarios where development of an international basin follows the logic of the Prisoner's Dilemma: if I take unilateral action while you don't, I can benefit (upper right and bottom left); however, if we both take unilateral action (bottom right), we are worse off than if we manage the resource jointly (top left). The lessons of the Prisoner's Dilemma suggest that open games and repeated games may help move the result toward cooperation. That is, we may be more likely to solve the collective action problem if activities are open and transparent (hence the importance of data sharing) and if both countries know that they are in this relationship for the long haul (hence the importance of stable regimes and institutions).

It should be noted that the Prisoner's Dilemma assumes that both players are in an equal position. In water resource disputes, however, one riparian usually is in a position of greater power, for one of several reasons: it is hydrologically upstream, it has military or political superiority, or it has other water resources and is thus less dependent on this particular shared water.

The game of Chicken may better represent some water situations. The payoff matrix for Chicken (Table 13.2) differs from Prisoner's Dilemma in that the worst outcome for both players is when both defect. Thus, the bottom right box is no longer a Nash equilibrium; given that one player is defecting, the other player can improve his outcome unilaterally by changing his action to "cooperate." Chicken has two Nash equilibria: the

TABLE 13.2
Payoff matrix for Chicken

Each player chooses whether to cooperate or defect, with player A's choices represented by the columns and player B's choices by the rows. Each of the four cells of the table then represents an outcome of the game, with the payoff to player A indicated by the number above the diagonal line and the payoff to player B indicated by the number below the diagonal line. Higher payoffs are better.

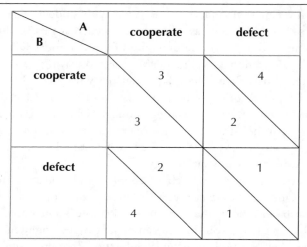

bottom left and top right corners, in which one player cooperates while the other defects. Chicken thus tends to exacerbate preexisting gaps between the two players: the one with greater power or greater willingness to take risks tends to defect, while the other player tends to cooperate.

The Pareto-optimal outcomes in Chicken are the same three boxes as in Prisoner's Dilemma. Thus, unlike Prisoner's Dilemma, Chicken does tend to lead to one of the Pareto-optimal outcomes (bottom left and top right). While these outcomes are Pareto-optimal, however, they are not equitable, and most observers would argue that the top left corner is more socially desirable. To the extent that water resource disputes take the form of a game of Chicken (at least for one player), it may be hard to move the result away from the inequitable but Pareto-optimal outcome.

## Water Conflict and Cooperation in the Jordan River Basin

To make these abstract principles a bit more concrete, let us examine the realities of hydropolitics in the Jordan River basin.

The Jordan is a relatively small river that has outsized importance culturally, economically, ecologically, and politically. Culturally, human civilization has ancient roots in the Jordan Valley, and the river figures prominently in the histories of Christianity, Islam, and Judaism. Economically, the Jordan's waters are a vital resource that is essential to the survival of local communities, including several million people outside its physical watershed. Ecologically, the Jordan occupies a distinctive position by virtue of its Mediterranean climate, its location on the Great Rift bird migration corridor, and its discharge to the Dead Sea. Politically, the Jordan is both a locus of tension and a potential source of cooperation in this conflict-ridden region.

Despite its relatively small area, the Jordan watershed includes parts of five political entities (Lebanon, Syria, Israel, Jordan, and the Palestinian Authority) and contains remarkably sharp climatic gradients, from mountainous headwaters (up to ~900 mm/yr of precipitation) to the hot, arid valley through which the river flows to its terminus in the Dead Sea. Rainfall occurs only in winter and is highly variable from year to year, with multiyear droughts a common occurrence. The vast majority of streamflow comes from two tributaries—the Upper Jordan (which flows into the Sea of Galilee) and the Yarmouk—each of which would naturally generate about 500–700 MCM in an average year.

The entities that depend on the waters of the Jordan—especially Jordan, the Palestinian Authority, and Israel—are among the most water-scarce in the world. As their populations and economies continue to grow, these entities are dealing with water scarcity by importing virtual water and depleting their groundwater reserves, in addition to fully utilizing the waters of the Jordan River.

Global-scale climate models are mostly in agreement that the Jordan basin is one of the locations where climate change is likely to lead to a decrease in runoff (Milly et al. 2005, Nohara et al. 2006). In addition, a high-resolution (20 km) climate model for this area suggests that streamflow generation in the Jordan basin could drop to near zero by the end of this century (Kitoh et al. 2008), a development that would have very severe consequences for the economies and health of the riparian countries. The reliability of this model has been questioned (Ben-Zvi and Givati 2008).

Experts disagree over the extent to which water is a driver of violent conflict in the region (Amery and Wolf 2000, Feitelson 2000, Amery 2002, Zeitoun et al. 2009), but it seems safe to note that water issues have not been the primary reason for the Arab-Israeli conflict. Likewise, it is an exaggeration to say—as some have claimed—that Israel's military and diplomatic strategy has been governed by a "hydraulic imperative":

a desire to control vital sources of water. In fact, water has in some cases been a leading issue in building relationships, as in the famous "Picnic Table Talks," in which Jordanian, and Israelis met periodically to coordinate their management of the river, even when their countries were technically at war. At the same time, it is clear that the development of the Jordan River's water resources over the last 50 years has been characterized largely by unilateral actions and threats of violence.

Israel, Syria, and Jordan each have developed their own infrastructure for capturing the Jordan basin's water. Israel uses its National Water Carrier, completed in 1964, to divert ~450 MCM/yr from the Sea of Galilee to the populated coastal plain; it also makes several smaller local withdrawals. Syria has built several dozen dams on tributaries of the Yarmouk and is now using some 200 MCM/yr from the basin. Jordan diverts about 175 MCM/yr of the Yarmouk flow to its King Abdullah Canal, completed in 1966, to support irrigation in the Jordan Valley; it also has other surface and groundwater withdrawals in the basin for other agricultural and urban needs. The 2008 al-Wahda (Unity) Dam on the Yarmouk was built jointly by Jordan and Syria, but Jordanian water officials attribute the slow filling of the reservoir to clandestine upstream withdrawals by Syria.

This unilateralism has had several negative consequences. First, it has led to threats of violence and actual military conflict, as in the 1960s when Israel and Syria exchanged fire both over Israel's work on the National Water Carrier and over Syria's attempts to divert the headwaters of the Upper Jordan. Second, there have been real economic and hydrologic costs to the lack of integrated management. The most dramatic example may be the location of the intake to the National Water Carrier, which was originally planned for the Upper Jordan, but was moved for geopolitical reasons to the Sea of Galilee, where salinities are higher and the elevation is lower, requiring huge energy investments in water pumping. The third impact of unilateralism has been the devastation of the Lower Jordan River ecosystem, which is due at least in part to a "tragedy of the commons" and to the view of this border river as an inaccessible danger zone, rather than a joint natural heritage. Remaining flows in the lower river consist mostly of salt water and agricultural return flow and are on the order of 5% of natural flow, except in years when high rainfall results in floods that can't be captured by existing infrastructure.

Given the complex territorial claims and different levels of development in the basin, what is a fair water allocation among the riparians? In the early 1950s, as tensions were escalating over separate Israeli and Arab plans for the river, the US sent Eric Johnston to the region to negotiate an agreement that would allow integrated development. The Johnston Plan was never formally accepted by all the parties but is generally considered the starting point for future water negotiations. Both Israel and Syria have been significantly overusing their Johnston allocations. Caught in the middle are the arid Kingdom of Jordan, which is hydrologically and politically weaker than its upstream neighbors, and the Palestinian villages of the West Bank, whose traditional water sources—small local springs and streams draining to the Jordan—have been affected by Israeli groundwater pumping. (The extensive groundwater resources of the West Bank constitute a separate, highly contested resource.)

One bright point is the Israeli-Jordanian peace treaty of 1994, which included a strong water component. The most important features of the water agreement include specification of the rights of each country in both the Yarmouk and the Lower Jordan (including Israeli compliance with the Johnston allocation on the Yarmouk), construction of infrastructure to transfer water between the two countries (including joint real-time monitoring of water quantity and quality), cooperation to store Jordanian water in Israel's Sea of Galilee, and a commitment to jointly find new sources of water (e.g., through desalination) to meet Jordanian needs. The water treaty was the result of intense and protracted negotiations, but the final agreement had significant areas of ambiguity, which have been open to different interpretations by the two sides. This ambiguity is both a strength and a weakness (Fischhendler 2008b): the former because it allowed each side to sell the treaty to its own public and the latter because of the inevitable disagreements that have surfaced in implementing the treaty (e.g., differences over the source of funding for the additional water to be provided to Jordan).

A second bright point is the emergence of governmental and nongovernmental groups working for cooperation over water, often with both regional and international involvement and funding. One such group, Friends of the Earth Middle East (FoEME), has been at the forefront of environmental peace making. FoEME sees the waters of the Jordan Basin as an opportunity to build relationships between local Israeli, Palestinian, and Jordanian communities for better water management, through projects such as Good Water Neighbors. It is working toward environmental rehabilitation of the river and toward reaching the day when the benefits of a healthy riverine ecosystem are shared among the different riparians.

## 13.2. Water Conflict and Cooperation in the US

While the US shares water resources with both Mexico and Canada—and has water treaties and conflicts with each—a great deal of the conflict over water in the US occurs between states or between users within a state. Many of the principles discussed above at the international scale apply here as well, though they often play out in different ways.

### The Law of the River

The legal basis for water allocation among different users within a state was discussed in section 12.2, where we summarized both the riparian doctrine that dominates in the eastern states and the prior appropriation system that holds in the western states. In practice, for large rivers, especially in the western states, the "law of the river" is quite complex and includes several elements beyond the simple individual water use rights discussed previously:

- Interstate water allocation: Where river basins or aquifers encompass multiple states, the amount of water allocated to each state must be determined. When water disputes arise between states, they can be resolved by the Supreme Court—which applies the doctrine of "equitable apportionment"—or by legislation. However, the most desirable outcome is generally an interstate compact (which must be approved by Congress) resulting from negotiations between the parties. Interstate compacts can specify water quality as well as the amount of flow allocated to each state and should ideally include a description of how to deal with shortfalls (e.g., by allocating percentages of total flows rather than absolute amounts).

- Federal reserved water rights: The *Winters doctrine* (based on the 1908 case of *Winters v. US*) states that when the federal government established Indian reservations and other protected lands (e.g., National Parks), it provided them with an implicit water right. This water right has a filing date identical to the date of establishment of the reservation, which usually makes it senior to most other water rights. While the doctrine is well established, the size of the water right involved is not. National Parks and tribes tend to argue for expansive estimates of the amount of water needed to fulfill the purposes of the reservation, while other users want to minimize the size of this water right.

- Poorly documented appropriations: Even for users within a state, water rights may not be that clear, especially when the state follows the prior appropriation doctrine—so that the filing date is the key to the seniority of one's water rights—and the records of amounts and dates are shrouded in the dust of the nineteenth century. Several western states have turned to *general stream adjudications* to sort out all the conflicting claims on the river (including federal reserved rights), but these have dragged on for decades.
- Endangered Species Act: As mentioned in section 8.7, the ESA has increasingly been used to require in-stream flows to support fish habitat, often at the expense of other users, even those with senior water rights. A key unsettled legal question is the extent to which the loss of these water allocations constitutes a "takings" of property, which must be compensated by the government. Many were surprised in 2001 by the "Tulare" decision, in which a judge ruled that farmers in California's Central Valley must be compensated for the losses that they suffered when the irrigation water for which they had long-term contracts was instead released into the Sacramento–San Joaquin Delta to protect the endangered delta smelt and winter-run chinook salmon. The Bush administration chose to settle the suit rather than appeal it, but other similar cases have resulted in opposite outcomes, and the matter will probably need to ultimately be settled by the Supreme Court.
- Clean Water Act: The CWA, while it deals primarily with water quality, can also have a significant impact on water allocation. One way in which this happens is through Section 404, which requires a permit from the Army Corps of Engineers (in consultation with other agencies, including EPA) for any dredging or filling operation in US waters. This means that construction of dams and other large water projects can be blocked by denying these permits.

How do all these different forces interact to produce conflict or cooperation in water management? The arid West has a long history of water conflict, exemplified by the battles over the Colorado River. However, these "water wars" have come east as well, as population growth, increased consumption, drought, and awareness of in-stream (environmental) water needs have combined to create a "perfect storm" of scarcity and conflict in certain river basins. We now examine two case studies that exemplify many of the issues involved.

## Klamath River

The Klamath River has been the site of arguably the most heated "fish vs. farmer" water conflict in the US. The Upper Klamath Basin, in southern Oregon and northern California, is a high desert plateau, with precipitation that is quite low (average ~350 mm/yr), seasonal (winter only), and highly variable (frequent drought). The lower basin, in northwestern California, is wetter and steeper than the upper basin, leading some to refer to the Klamath as an "upside-down" watershed (Doremus and Tarlock 2008).

The Klamath Basin is inhabited by a complex group of actors with different water needs and perspectives (Doremus and Tarlock 2008):

- Farmers: The main water users in the upper basin are ~1700 irrigated farms, mostly small family operations, growing potatoes, alfalfa, hay, and pasture. Their economic survival is completely tied to the availability of irrigation water, which they obtain mostly from the Bureau of Reclamation under long-term contracts dating back decades.
- Bureau of Reclamation (BuRec): One of BuRec's first projects, the Klamath Project, begun in 1906, is a complex system of dams for water storage and canals for water delivery. Water storage, which is primarily in Upper Klamath Lake, is only seasonal, not interannual, so the system does not provide much protection against drought. BuRec has senior appropriative rights to most of the water of the upper basin.
- Hydropower: The lower part of the upper basin contains five hydroelectric dams, operated by PacifiCorp and providing cheap electricity to upper basin farmers and the larger region. Hydropower water rights in the basin are generally junior to irrigation rights.
- Wildlife: Several National Wildlife Refuges have been established in the upper basin, which is home to large populations of wildlife and birds attracted by the lakes and marshes of the region. Habitat in the refuges is threatened by excessive water withdrawals, land drainage, and polluted return flows. In theory, the refuges have reserved water rights (Winters rights), but these have not been quantified.

- Fish: Two species of sucker were once prolific in the streams and lakes of the upper basin but are now listed under the ESA. In addition, several species of Pacific salmon migrate between the ocean and the tributaries of the Klamath—or did, until dams, water withdrawals, and overfishing depleted their numbers. Coho salmon runs are now listed under the ESA, and chinook salmon have also declined dramatically.
- Fishers: Both commercial ocean fishers and river sports fishers rely on the salmon runs. Severe fishing restrictions have been placed on the regional commercial salmon fishery in order to protect the Klamath chinook, resulting in significant economic losses.
- Native tribes: The basin is home to five different American Indian tribes that traditionally lived primarily off fish (suckers in the upper basin, salmon in the lower basin) and now claim their reserved water rights for in-stream flows to protect the fish.

This mixture of apparently irreconcilable demands reached a flash point in 2001. A BiOp (biological opinion; see section 8.7) from the Fish and Wildlife Service had required a minimum water level in Upper Klamath Lake to protect the suckers, while a BiOp from NOAA Fisheries had required a minimum flow into the lower basin to protect the coho. After a dry winter, there simply was not enough water to meet these requirements and also provide irrigation water, and in April 2001 BuRec announced that it would not release any water to farmers. The uproar that ensued included mass protests, civil disobedience, vandalism to canal infrastructure, and a $1 billion takings lawsuit (unsuccessful). In 2002, apparently under pressure from the Bush administration, a revised BiOp was published that allowed water to be released for irrigation. This resulted in a large fish kill downstream, and a judge later rejected the revised BiOp.

Since 2002, tempers have calmed a bit, and a solution to at least one part of the basin's problems is emerging, in the form of an agreement to remove four of the hydroelectric dams in order to restore salmon runs. It should be noted, however, that this agreement does little to resolve the other basic problem facing the basin: insufficient water to meet the demands of both agriculture and the environment.

To many, the Klamath situation represents a struggle between the old West of irrigation and hydropower and a new West of environmental protection and tribal rights. Is there a way to reconcile these different demands on the resource, or is the old West destined to fall before the power of the ESA and the Winters doctrine?

## Apalachicola-Chattahoochee-Flint (ACF)

A "Western-style" interstate water war has been raging for years among the southeastern states of Georgia, Alabama, and Florida over the waters of the ACF river system. The Chattahoochee and Flint rivers originate in Georgia, flow through Alabama, and merge near the Florida border to form the Apalachicola River, which discharges into Florida's Apalachicola Bay on the Gulf of Mexico. (A second water war over the adjacent Alabama-Coosa-Tallapoosa [ACT] river system also involves Georgia and Alabama but isn't discussed here.)

The players in the ACF conflict include the following:

- Atlanta: Despite being blessed with high rainfall, the metropolis of Atlanta is in a vulnerable hydrologic position. For one thing, it is located near the headwaters of several river systems, which means that it only has available to it runoff from quite a small area. In addition, the region is susceptible to periodic droughts—such as the one experienced in 2006–2008—and has relatively little groundwater. Add to this a growth rate that is among the highest in the country, as well as high per capita water use, and you have a recipe for water scarcity. Over the last several years, Atlanta has relied on increasing withdrawals from Lake Lanier, a reservoir on the Chattahoochee, to meet its water needs.
- US Army Corps of Engineers (ACOE): The ACOE is responsible for managing water levels and flows in the ACF system, including Lake Lanier and four other downstream lakes. The ACOE tries to balance the different uses of these systems, including hydropower (which requires release of water at certain times to produce peaking power), navigation (which requires continuous release of water to maintain a 9-foot depth of water in the channel), recreation (which generally is improved by keeping lake levels high), flood control (which requires releasing water soon after a flood event in order to prepare for the next flood), water quality control (which requires releasing water to dilute downstream pollution), fish and wildlife (both in the lakes and downstream), and water supply (e.g., for Atlanta).
- Southern Georgia farmers: Farmers in the Flint River basin rely heavily on surface and groundwater withdrawals for irrigation.

Despite being part of the same state as Atlanta, they comprise a distinct interest group.

- Alabama: Alabama's main interests in the river are navigation, hydropower, and water supply for industrial, agricultural, and domestic use.

- Florida: Florida has pushed for maintaining adequate delivery of freshwater to Apalachicola Bay, necessary for maintaining the ecological health of the oysters that are the foundation of an industry worth more than $60 million per year.

The battle among these conflicting interests began in 1990, with a lawsuit by Alabama (joined by Florida) against Georgia's plan to withdraw more water from Lake Lanier. After years of failed negotiations and inconclusive court battles, a decisive point in the process was reached recently, when a federal court ruled that the ACOE had no right to change the rules governing allocation of water from Lake Lanier. This leaves Atlanta as the clear loser and as a city in desperate need of water solutions.

Will a compromise be reached to accommodate the needs of this powerful—but resented—urban center? Will the legal battle continue toward the Supreme Court and an "equitable apportionment"? Will Atlanta limit its growth to stay within its water constraints? Will serious conservation programs free up enough water to alleviate the need for additional supply? Or will the next drought once again find politicians praying for rain and fighting over reservoir releases?

## 13.3 Conclusion

In this chapter, we examined how different countries, states, and users share water resources. At the international scale, rules governing water allocation are weak, but the evidence suggests that cooperation is more common than conflict—although there are certainly instances of the latter. At the subnational scale (in particular, within the US), rules are much stronger, but the complexity of these rules, along with changing socioeconomic, cultural, and environmental conditions, make legal conflict common.

In the coming decades, increasing water scarcity and environmental degradation are likely to strain our capacity to share water resources at both the international and subnational levels. We will have a great need for strong institutions, clear but flexible rules, creative thinking, and goodwill in order to work out arrangements for balancing different needs and jointly managing watersheds.

# 14

# Conclusion: The Imperative of Better Management

We began this book with an overview of the water crisis (section 1.2), divided into 10 components, corresponding to Chapters 4–13. In the pages that have followed, we have delved into the details of this crisis and have explored the complex issues involved. We have also looked at some tools—technical, economic, legal, social—that could be used to manage water better and help move us toward solutions.

We close by returning to the same 10 components and briefly summarizing some of what we need to do to ease the water crisis.

Flooding (Ch. 4). We need to work with, rather than against, the power of water, in order to minimize the damage done by floods. We need to reduce our vulnerability to flood events and take action in our floodplains and throughout our watersheds to restore their natural ability to absorb water.

Scarcity (Ch. 5). We need to use both new and old tools to ease water scarcity: reducing pollution discharges to protect natural water sources, increasing water supply through judicious use of appropriate technologies, reducing water demand in all sectors through conservation and increased efficiency, and reallocating water to more efficient, higher-valued uses.

Change (Ch. 6). We need to be prepared for substantial changes in water supply and demand. Reducing greenhouse gas emissions and slowing the pace of land use change can help us avoid some of the worst changes to the water cycle, while building flexibility into our water systems can help us deal with the changes that do come.

Technology (Ch. 7). We need to expand our thinking beyond the technologies of the twentieth century to include traditional and emerging technologies like rainwater harvesting, wastewater reclamation, desalination, virtual water trade, and conservation. While new dams and canals still should be built in some places, we need to expand our horizons and improve our ability to evaluate which approaches are best suited to particular situations.

Ecosystem degradation (Ch. 8). We need to better balance human and ecological water use and improve our protection and restoration of aquatic ecosystems. Most fundamentally, our decision-making process needs to fully take into account the consequences of our actions for ecosystems and natural infrastructure.

Human health (Ch. 9). We need to provide basic water and sanitation services to billions of people who lack them, while also improving the safety of the water supply throughout the world. We need to reduce leaks and waste and implement water-efficient household technologies.

Agriculture (Ch. 10). We need to make more efficient use of water in both rain-fed and irrigated agriculture by increasing yields (especially in poorer countries), implementing new irrigation technologies, removing subsidies, and reducing meat consumption. We also need to reduce the impact of agricultural pollution on water resources and aquatic ecosystems.

Industry (Ch. 11). We need to reduce water use in energy production and in manufacturing and minimize the pollution caused by these activities.

Inefficiency and inequity (Ch. 12). We need to use economic tools selectively and wisely to improve the efficiency and equity of water allocation and water use. In many situations, we can improve water management through better pricing policies, market-based reallocation of water to higher-valued uses, and the involvement of private capital and expertise in water supply and sanitation.

Conflict (Ch. 13). We need to minimize the potential for water conflict by providing clear but flexible water rules and building strong transboundary institutions and partnerships.

If we can do these things—and I believe that we can—we will secure a better water future for our children and for the entire planet.

# Discussion Questions

## Chapter 1

1. Which aspects of the water crisis do you have the most experience with? Which do you think are most important?
2. What are the most important characteristics of the emerging approach to water management?

## Chapter 2

1. What are the advantages and disadvantages of the watershed as a unit of science and management?
2. In what ways is an understanding of hydrology and water quality important for water resource managers?
3. How large is the watershed you live in? How does this fact affect the ways that water is (or should be) managed?
4. Is there abundant groundwater in your area? How is it connected to surface water? How do these facts affect the ways that water is (or should be) managed?

## Chapter 3

1. What do you think of the usefulness and feasibility of assessing TRWR at the country scale?
2. Should water conservation efforts focus primarily on reducing consumptive uses, or should nonconsumptive uses also be addressed?
3. Does anything in the data on country-level water use and water footprints surprise you?
4. What lessons do you draw from the decline in water withdrawals in the US over the last several decades?

## Chapter 4

1. Is the 100-year flood a useful concept?
2. What are the strengths and weaknesses of structural flood control?
3. Do you agree with the three principles proposed at the end of section 4.3?

## Chapter 5

1. What are the strengths and weaknesses of the different water scarcity indicators?
2. What is the most appropriate scale for assessing water scarcity?
3. When is it appropriate to use groundwater at a rate faster than the rate of recharge?
4. Does increasing water use efficiency make water systems more or less vulnerable to drought?

## Chapter 6

1. What is the relative importance of climate, population, and land use as drivers of changing water stress in the area where you live?
2. How can water management become more resilient to change?
3. What is your perspective on the scientific uncertainty associated with predicting the impacts of climate change on water resources? Are the data and models good enough for managers to act on?

## Chapter 7

1. How does one evaluate which technologies are best suited for water management in a particular situation?
2. What are the advantages and disadvantages of dams, water transfers, virtual water transfers, desalination, wastewater reclamation, and rainwater harvesting? Which of these technologies do you see as most necessary for future water management?

## Chapter 8

1. Does Smakhtin et al. (2004)'s simple approach to calculating environmental flow requirements do more harm than good (as argued by Arthington et al. 2006)? Or is it important to have a basic tool that can be applied globally without exhaustive study of each individual river?
2. In this chapter, we discussed four types of impacts to aquatic ecosystems: hydrologic, geomorphic, chemical, and biotic. What are some of the relationships among these different categories? Which category do you think is most important? Which seems most understudied?
3. Based on the snapshot of river health provided by EPA's Integrated Report database, do you think that we are doing a good enough job at balancing human and ecosystem needs in the US?

4. What are some of the problems with the CWA approach to assessing ecosystem health?

## Chapter 9

1. Do you see drinking water safety in the US as a glass half full or half empty? What are the strengths and weaknesses in our drinking water supply system?
2. What aspects (if any) of the "standard model" for drinking water and sanitation could and should be replicated worldwide?
3. Is POU treatment a valuable tool for improving health outcomes or is it a distraction from meeting the MDGs for water and sanitation access?
4. What are the most effective tools for household water conservation?

## Chapter 10

1. What are the strengths and weaknesses of different definitions of agricultural water use efficiency?
2. What are the highest-priority actions we can take for reducing agricultural water footprints?
3. How can we better manage agricultural pollution?

## Chapter 11

1. Is the shift from once-through to recirculating cooling a good thing?
2. Is hydropower truly a renewable, emission-free source of energy?
3. How effective has TRI been at reducing industrial pollution?

## Chapter 12

1. What are the strengths and weaknesses of each of these approaches to water allocation: riparian rights, prior appropriation, regulated riparianism, and water markets?
2. In your view, what is the appropriate role for private companies in water and sanitation provision?
3. In your view, how should water prices be set?

## Chapter 13

1. What are the most important tools for moving international water conflicts toward cooperation? What about for subnational water conflicts?

2. What lessons for water management do you take from the Klamath and ACF cases?
3. How does game theory help us understand international water conflicts?

# Glossary

ACOE (Army Corps of Engineers): a US government agency that manages many dams and water projects, especially in the East

agricultural water use: water used in growing crops and raising livestock; generally includes conveyance losses

anthropogenic: caused by humans

AQUASTAT: a global database on water use and availability, maintained by FAO

BiOp (Biological Opinion): an assessment (carried out by either FWS or NOAA Fisheries) to determine whether a proposed action would jeopardize the continued survival of an endangered species; required by the ESA

blue water: surface or groundwater; a water resource that can be extracted and used by people

BMP (best management practice): used in this book to mean structural stormwater best management practice, a structure designed to slow down the movement of stormwater to the stream and, in some cases, also provide water quality improvement

BOD (biochemical oxygen demand): the amount of oxygen that would be consumed during respiration of the organic material in a water body or water sample

BuRec (Bureau of Reclamation): a US government agency (part of the Department of the Interior) that manages many dams and water projects in the West

CAFO (concentrated animal feeding operation): a feedlot at which large numbers of animals are being raised without access to pasture

CBA (cost-benefit analysis): an economic analysis of the costs and benefits of a potential project, used for determining whether to move forward with the project

cereal crop: a grass grown for its fruit seed, such as corn, rice, wheat, and barley

cfs: cubic feet per second, $ft^3/sec$, a measure of flow (see Box 3.1)

closed basin: a watershed where the entirety of runoff is allocated to human use

consumptive water use: the amount of water that is consumed (e.g., evaporates) during use conveyance losses: water lost to seepage or evaporation between withdrawal from a source and delivery to a user (especially used in the agricultural context)

CSO (combined sewer overflow): an event in which a combined sewer (one carrying both urban runoff and sewage) exceeds its capacity during a rainfall event and overflows to a nearby water body

CV (coefficient of variation): statistical measure of variability; the standard deviation divided by the mean (often expressed as a percent)

CWA (Clean Water Act): US legislation, meant to protect and restore aquatic ecosystems

demand management: an approach to dealing with a potential imbalance between water supply and demand, by taking various measures to reduce demand

desalination: the removal of salts from brackish or salt water to produce water that can be used for domestic, agricultural, or industrial uses; also called desalinization or desalting

DO (dissolved oxygen): the amount of oxygen dissolved in a water body or water sample

domestic water use: water used by households; also generally includes water provided by utilities for use by institutions, commercial establishments, and small industrial facilities; also generally includes the utility's unaccounted-for water

drought: unusually dry conditions, reflected in low precipitation (meteorological drought), low soil moisture for crops (agricultural drought), low surface water or groundwater levels (hydrologic drought), and/or resultant impacts on human society (socioeconomic drought)

EIA (effective impervious area): the portion of the TIA that is directly connected to water bodies through a network of impervious area and storm sewers

endorheic region: a watershed that drains to an inland lake, sea, or sink, from which there is no outflow to the ocean

environmental flows: water moving in a stream or river (as opposed to withdrawn for human use); also refers to the amount of such water flow necessary for sustaining aquatic ecosystems

EPA (Environmental Protection Agency): a US government agency responsible (among other things) for implementing the CWA and SDWA

ESA (Endangered Species Act): US legislation, meant to protect endangered species from extinction

ET (evapotranspiration): movement of water to the atmosphere from terrestrial or aquatic ecosystems through the combination of evaporation (physical-chemical movement of water from liquid to gas state) and transpiration (release of water to the atmosphere by plants)

EWR (environmental water requirement): the amount of flow necessary to sustain the health of aquatic ecosystems, especially as calculated by Smakhtin et al. (2004)

Falkenmark indicator: a scarcity indicator, defined as the level of per capita water availability

FAO (Food and Agriculture Organization): United Nations organization responsible (among other things) for collecting global data on agriculture and water use

FAOSTAT: a global database on agricultural productivity and land use, maintained by FAO

FERC (Federal Energy Regulatory Commission): a US government agency responsible (among other things) for licensing hydroelectric power plants

FIRM (Flood Insurance Rate Map): maps produced by the NFIP that delineate different levels of flooding probability, corresponding to different regulations on land use

fossil groundwater: groundwater that was deposited over many millennia or under climatic conditions different from the present and is no longer being recharged

FWS (Fish & Wildlife Service): a US government agency (part of the Department of Interior), responsible (among other things) for consulting on actions affecting endangered freshwater species

GDP (Gross Domestic Product): a measure of a country's total economic output; per capita GDP is often used as a measure of a country's level of economic development

GHG (greenhouse gas): a gas that contributes to the warming of Earth through the greenhouse effect; the main anthropogenic greenhouse gases are carbon dioxide and methane

graywater: lightly used water (e.g., from dishwashing, laundry, bathing) that can be used again (e.g., for watering plants); can also be used (in the water footprint context) to mean the amount of water needed to dilute the pollution associated with a particular activity

green water: soil water; a water resource that is used by plants (including crops) for growth but can't be extracted and used elsewhere by people; can also be used to mean the water evaporated in growing something, even if that water originated as blue water

groundwater mining: use of groundwater at a rate higher than it is being recharged, leading to a lowering of the water table

groundwater overdraft: use of groundwater at a rate higher than it is being recharged, leading to a lowering of the water table

groundwater recharge: movement of water from the surface down into the groundwater zone

hard path: the dominant global approach to water management in the twentieth century, with an emphasis on large, centralized infrastructure and increasing water supply

HUC (hydrologic unit code): a geographic unit used by the USGS for managing water data; many, but not all hydrologic units are watersheds

industrial water use: water used by industrial facilities, including that used in energy production

in-stream flow: water moving in a stream or river (as opposed to withdrawn for human use); also refers to the amount of such water flow necessary for sustaining aquatic ecosystems

in-stream water use: water use that does not involve removing water from a water body; includes navigation, most hydropower, recreation, and environmental uses

interbasin water transfer: a transfer of water in which the supplying and receiving water bodies are in different watersheds

ITCZ (intertropical convergence zone): a band of high precipitation near the equator, caused by the convergence of surface air masses, leading to rising air and adiabatic cooling

IWRM (integrated water resource management): an emerging approach to water management, with an emphasis on inclusion of all stakeholders, strong governance, and the balancing of efficiency, equity, and environmental sustainability

large dam: a dam that meets one of the following criteria: height >15 m or reservoir volume >3 MCM

levee: a wall-like structure built to constrain and control water flow; often built along riverbanks to prevent the river from overflowing during floods

major dam: a dam that meets one of the following criteria: height >150 m, reservoir volume >25,000 MCM, dam volume >15 MCM, or installed hydroelectric capacity >1000 megawatts

MCM: million cubic meters (see Box 3.1)

MENA: Middle East and North Africa

MNB (marginal net benefits): the net benefits (benefits minus costs) obtained from the next increment of resource use

N (nitrogen): a nutrient element necessary for plant growth

Nash equilibrium: outcome (of a game or conflict) in which neither party can act unilaterally to improve its own position

NFIP (National Flood Insurance Program): a program designed to reduce flood losses by providing flood insurance while also encouraging sensible use of the floodplain

NOAA (National Oceanographic & Atmospheric Administration): a US government agency (part of the Department of Commerce), responsible (among other things) for consulting on actions affecting endangered marine species

nonconsumptive water use: the amount of water that is not consumed during use and returns to the local environment

NPDES (National Pollutant Discharge Elimination System): a program established by the CWA that requires permits for all point sources of water pollution

NPS (nonpoint source): a diffuse source of pollution, such as a large land area that contributes polluted runoff

nutrients: elements that are necessary for production by plants or algae, especially nitrogen and phosphorus

OECD (Organisation for Economic Co-operation and Development): international organization of 30 industrialized countries

off-stream water use: water use that involves withdrawing water from a water body

P (phosphorus): a nutrient element necessary for plant growth

Pareto-optimal outcome: outcome (of a game or conflict) that meets the following criterion: no other outcome exists that is better for one party without being worse for another party

PET (potential evapotranspiration): the amount of evapotranspiration that would occur from a well-watered surface

point source: a localized source of pollution, such as a factory

prior allocation: a water rights system in which water is allocated based on the seniority of claims

Q90: the flow in a river that is exceeded 90% of the time; can be calculated at different time scales (e.g., the daily flow that is exceeded 9 out of 10 days, the monthly flow that is exceed 9 out of 10 months, etc.)

rainwater harvesting: a diverse group of techniques used to capture rainfall for local use

reclamation: the treatment of wastewater to the level where it can be reused; also can be used to mean conversion of natural ecosystems to human use

recurrence interval: a statistical measure of the probability of a given event; equal to one over the annual probability of occurrence (see Box 4.1)

riparian: a person, state, or country that owns land that touches a water body

riparian rights: a water rights system in which water is shared among all riparians to the water body

runoff: water that flows out of a watershed as surface water (generally in a river)

SDWA (Safe Drinking Water Act): US legislation, meant to ensure the safety of drinking water

seepage: water that infiltrates the soil, especially during movement through a canal

soft path: an emerging approach to water management, with an emphasis on decentralized projects, preserving ecosystems, and reducing demand for water

stationarity: the idea that water resources fluctuate within a range that is unchanging over time

STP: sewage treatment plant

subsidence: sinking of the land surface, often due to groundwater withdrawal or wetland drainage

supply management: an approach to dealing with a potential imbalance between water supply and demand, by taking various measures to increase supply

TIA (total impervious area): the fraction of a watershed (usually expressed in percent) that is not pervious to water infiltration (e.g., pavements, rooftops)

TMDL (total maximum daily load): the maximum amount of a pollutant that a water body can receive while still meeting WQC; also the document that calculates this amount and describes how it will be achieved

TRI (Toxics Release Inventory): a database of self-reported industrial use and release of toxic chemicals, required by the Emergency Planning and Community Right-to-Know Act of 1986

TRWR (total renewable water resources): the average amount of renewable water flow available per year in a given country or other unit; includes surface water and groundwater recharge but not groundwater overdraft

USGS (US Geological Survey): a US government agency (part of the Department of the Interior), responsible (among other things) for research and data-gathering on surface water, groundwater, and water use

virtual water: the water that was used to grow a food or produce a product

water consumption: the amount of water that is consumed (e.g., evaporates) during use

water footprint: the amount of water required to produce a product or grow a crop; also the total amount of water required for all the products or crops used by a person or country

watershed: the topographic area within which apparent surface water runoff drains to a specific point on a stream or to a water body such as a lake

watershed approach: EPA's approach to IWRM, with an emphasis on stakeholder inclusion, use of science, and geographically defined integrated management

water table: the elevation of the top of the groundwater zone

water withdrawal: the amount of water that is withdrawn from a water body for off-stream use; includes both consumptive and nonconsumptive water use

WCD (World Commission on Dams): an international group that studied the costs and benefits of large dams and produced an analysis and recommendations in 2000

WHO (World Health Organization): UN organization, responsible (among other things) for tracking water/sanitation access and water-related disease

Winters doctrine: a US water rights doctrine stating that federal land reservations (such as National Parks or Indian reservations) have an implicit water right

WQC (water quality criteria): acceptable levels of different pollutants in water bodies, established under the CWA with the goal of protecting human and ecosystem health

WTA (withdrawal-to-availability) indicator: a scarcity indicator, defined as the percent of available water that is currently being withdrawn

# References

Albert, J. 2000. Rethinking the management of transboundary freshwater resources: A critical examination of modern international law and practice. *Natural Resources Forum* 24:21–30.

Alcamo, J., L. Acosta-Michlik, A. Carius, F. Eierdanz, R. Klein, D. Kromker, and D. Tanzler. 2008. A new approach to quantifying and comparing vulnerability to drought. *Regional Environmental Change* 8:137–149.

Alcamo, J., P. Döll, T. Henrichs, F. Kaspar, B. Lehner, T. Rosch, and S. Siebert. 2003. Global estimates of water withdrawals and availability under current and future "business-as-usual" conditions. *Hydrological Sciences Journal—Journal Des Sciences Hydrologiques* 48:339–348.

Alcamo, J., M. Florke, and M. Marker. 2007. Future long-term changes in global water resources driven by socio-economic and climatic changes. *Hydrological Sciences Journal—Journal Des Sciences Hydrologiques* 52:247–275.

Allan, J. A. 1998. Virtual water: A strategic resource global solutions to regional deficits. *Ground Water* 36:545–546.

Allan, J. A. 2001. *The Middle East Water Question: Hydropolitics and the Global Economy.* London: I B Tauris and Co.

American Institutes for Research and NFIP Evaluation Final Report Working Group (AIR/NFIP). 2006. *The Evaluation of the National Flood Insurance Program—Final Report*, available at www.fema.gov/library/viewRecord.do?id=2573.

American Rivers, Friends of the Earth, and Trout Unlimited. 1999. *Dam Removal Success Stories: Restoring Rivers Through Selective Removal of Dams That Don't Make Sense*, available at www.americanrivers.org/library/reports-publications/dam-removal-success-stories.html.

American Society of Civil Engineers (ASCE). 2009. *2009 Report Card for America's Infrastructure.* Reston, VA: American Society of Civil Engineers.

Amery, H. A. 2002. Water wars in the Middle East: A looming threat. *Geographical Journal* 168:313–323.

Amery, H. A., and A. T. Wolf. 2000. *Water in the Middle East: A Geography of Peace.* Austin: The University of Texas Press.

AQUASTAT database of the Food and Agriculture Organization (FAO). 2009. www. fao.org/nr/water/aquastat/dbase/index.stm.

Arthington, A. H., S. E. Bunn, N. L. Poff, and R. J. Naiman. 2006. The challenge of providing environmental flow rules to sustain river ecosystems. *Ecological Applications* 16:1311–1318.

Associated Press. 2009. Destruction of Kenya's forest feeds deadly drought. *New York Times.* October 19, 2009.

Atlas, M. 2007. TRI to communicate: Public knowledge of the federal Toxics Release Inventory. *Social Science Quarterly* 88:555–572.

Ayoob, S., and A. K. Gupta. 2006. Fluoride in drinking water: A review on the status and stress effects. *Critical Reviews in Environmental Science and Technology* 36:433–487.

Barbier, E. B. 2003. Upstream dams and downstream water allocation: The case of the Hadejia-Jama'are floodplain, northern Nigeria. *Water Resources Research* 39:10.1029/2003wr002249.

Barlow, M. 2007. *Blue Covenant: The Global Water Crisis and the Coming Battle for the Right to Water.* New York: The New Press.

Barnett, T. P., J. C. Adam, and D. P. Lettenmaier. 2005. Potential impacts of a warming climate on water availability in snow-dominated regions. *Nature* 438:303–309.

Barringer, F. 2009. In California, desalination of seawater as a test case. *New York Times.* May 15, 2009.

Bauer, C. J. 2004. Results of Chilean water markets: Empirical research since 1990. *Water Resources Research* 40:11.

Bednarek, A. T. 2001. Undamming rivers: A review of the ecological impacts of dam removal. *Environmental Management* 27:803–814.

Ben-Zvi, A., and A. Givati. 2008. Comment on "First super-high-resolution model projection that the ancient 'Fertile Crescent' will disappear this century." *Hydrological Research Letters* 2:45.

Bhattarai, M. 2004. *Irrigation Kuznets Curve, Governance and Dynamics of Irrigation Development: A Global Cross-Country Analysis from 1972 to 1991, Research Report 78.* Colombo: International Water Management Institute.

Bjornlund, H. 2003. Efficient water market mechanisms to cope with water scarcity. *International Journal of Water Resources Development* 19:553–567.

Booker, J. F., A. M. Michelsen, and F. A. Ward. 2005. Economic impact of alternative policy responses to prolonged and severe drought in the Rio Grande Basin. *Water Resources Research* 41:10.1029/2004wr003486.

Booth, D. B., and C. R. Jackson. 1997. Urbanization of aquatic systems: Degradation thresholds, stormwater detection, and the limits of mitigation. *Journal of the American Water Resources Association* 33:1077–1090.

Brewer, J., R. Glennon, A. Ker, and G. Libecap. 2008. 2006 Presidential Address— Water markets in the West: Prices, trading, and contractual forms. *Economic Inquiry* 46:91–112.

Briscoe, J., and R. P. S. Malik. 2006. *India's Water Economy: Bracing for a Turbulent Future.* Oxford: Oxford University Press.

Brooks, D. B. 1993. Adjusting the flow—2 comments on the Middle-East water crisis. *Water International* 18:35–39.

Brooks, R., and E. Harris. 2008. Efficiency gains from water markets: Empirical analysis of Watermove in Australia. *Agricultural Water Management* 95:391–399.

Brouwer, R., S. Akter, L. Brander, and E. Haque. 2007. Socioeconomic vulnerability and adaptation to environmental risk: A case study of climate change and flooding in Bangladesh. *Risk Analysis* 27:313–326.

Brown, C., and U. Lall. 2006. Water and economic development: The role of variability and a framework for resilience. *Natural Resources Forum* 30:306–317.

Bruijnizeel, L. A. 2004. Hydrological functions of tropical forests: Not seeing the soil for the trees? *Agriculture, Ecosystems and Environment* 104:185–228.

Calder, I. R. 1999. *The Blue Revolution: Land Use and Integrated Water Resources Management*. London: Earthscan.

Cappiella, K., and K. Brown. 2001. *Impervious Cover and Land Use in the Chesapeake Bay Watershed*. Ellicott City, MD: Center for Watershed Protection.

Caspary, G. 2007. Institutional Incoherence in Development Policy? The Case of Environmental and Social Safeguard Systems in OECD-country Public Financing for Large Dams in Developing Countries, Insitut d'Etudes Politiques de Paris. Ph.D. thesis.

California Energy Commission (CEC). 2005. *California's Water-Energy Relationship*, CEC-700-2005-011-SF, available at www.energy.ca.gov/2005publications/CEC-700-2005-011/CEC-700-2005-011-SF.pdf.

Chakraborti, D., M. M. Rahman, K. Paul, U. K. Chowdhury, M. K. Sengupta, D. Lodh, C. R. Chanda, K. C. Saha, and S. C. Mukherjee. 2002. Arsenic calamity in the Indian subcontinent: What lessons have been learned? *Talanta* 58:3–22.

Chao, B. F. 1995. Anthropogenic impact on global geodynamics due to reservoir water impoundment. *Geophysical Research Letters* 22:3529–3532.

Chao, B. F., Y. H. Wu, and Y. S. Li. 2008. Impact of artificial reservoir water impoundment on global sea level. *Science* 320:212–214.

Chapagain, A. K., and A. Y. Hoekstra. 2004. *Water Footprints of Nations, Report 16*. Delft: UNESCO.

Chappell, A., and C. T. Agnew. 2008. How certain is desiccation in West African Sahel rainfall (1930–1990)? *Journal of Geophysical Research—Atmospheres* 113:20.

Charney, J., P. H. Stone, and W. J. Quirk. 1975. Drought in Sahara—Biogeophysical feedback mechanism. *Science* 187:434–435.

Chowdhury, M. R., and M. N. Ward. 2007. Seasonal flooding in Bangladesh—Variability and predictability. *Hydrological Processes* 21:335–347.10.1002/hyp.6236.

Coe, M. T., M. H. Costa, and B. S. Soares-Filhoc. 2009. The influence of historical and potential future deforestation on the stream flow of the Amazon River—Land surface processes and atmospheric feedbacks. *Journal of Hydrology* 369:165–174.

Colmenares, L. B., and M. D. Zoback. 2007. Hydraulic fracturing and wellbore completion of coalbed methane wells in the Powder River Basin, Wyoming: Implications for water and gas production. *AAPG Bulletin* 91:51–67.

Comprehensive Assessment of Water Management in Agriculture (CAWMA). 2007. *Water for Food, Water for Life: A Comprehensive Assessment of Water Management in Agriculture*. London and Colombo: Earthscan and International Water Management Institute.

Connecticut Department of Environmental Protection (CT DEP). 2007. *A Total Maximum Daily Load Analysis for Eagleville Brook, Mansfield, CT*. Hartford, CT.

Cooley, H., and P. H. Gleick. 2009. Urban water use efficiencies: Lessons from United States cities. Pp. 101–126 in P. H. Gleick, ed. *The World's Water 2008–2009: The Biennial Report on Freshwater Resources*. Washington: Island Press.

Cooley, H., T. Hutchins-Cabibi, M. Cohen, P. H. Gleick, and M. Heberger. 2007. *Hidden Oasis: Water Conservation and Efficiency in Las Vegas*. Oakland: Pacific Institute.

Costanza, R., W. J. Mitsch, and J. W. Day Jr. 2006. A new vision for New Orleans and the Mississippi Delta: Applying ecological economics and ecological engineering. *Frontiers in Ecology and the Environment* 4:465–472.

Dahl, Thomas E. 1990. Wetlands losses in the United States 1780's to 1980's. U.S.

Department of the Interior, Fish and Wildlife Service, Washington, DC. Jamestown, ND: Northern Prairie Wildlife Research Center Online.

Dai, A. G., P. J. Lamb, K. E. Trenberth, M. Hulme, P. D. Jones, and P. P. Xie. 2004. The recent Sahel drought is real. *International Journal of Climatology* 24:1323–1331.

Dai, A. G., T. T. Qian, K. E. Trenberth, and J. D. Milliman. 2009. Changes in Continental Freshwater Discharge from 1948 to 2004. *Journal of Climate* 22:2773–2792.

Dai, A. G., and K. E. Trenberth. 2002. Estimates of freshwater discharge from continents: Latitudinal and seasonal variations. *Journal of Hydrometeorology* 3:660–687.

Davis, L. W. 2008. Durable goods and residential demand for energy and water: Evidence from a field trial. *Rand Journal of Economics* 39:530–546.

Day, J. W., G. P. Shaffer, L. D. Britsch, D. J. Reed, S. R. Hawes, and D. Cahoon. 2000. Pattern and process of land loss in the Mississippi Delta: A spatial and temporal analysis of wetland habitat change. *Estuaries* 23:425–438.

de Fraiture, C., X. Cai, U. Amarasinghe, M. Rosegrant, and D. Molden. 2004. *Does International Cereal Trade Save Water? The Impact of Virtual Water Trade on Global Water Use.* Colombo: International Water Management Institute.

Dellapenna, J. W. 2007. Transboundary water sharing and the need for public management. *Journal of Water Resources Planning and Management—Asce* 133:397–404.

Delmas, R., C. Galy-Lacaux, and S. Richard. 2001. Emissions of greenhouse gases from the tropical hydroelectric reservoir of Petit Saut (French Guiana) compared with emissions from thermal alternatives. *Global Biogeochemical Cycles* 15:993–1003.

DeSimone, L. A. 2009. Quality of water from domestic wells in principal aquifers of the United States, 1991–2004: U.S. Geological Survey Scientific Investigations Report 2008–5227, 139 pp., available online at http://pubs.usgs.gov/sir/2008/5227.

Dinar, S. 2009. Scarcity and cooperation along international rivers. *Global Environmental Politics* 9:109-+.

Döll, P. 2009. Vulnerability to the impact of climate change on renewable groundwater resources: A global-scale assessment. *Environmental Research Letters* 4:035006.

Döll, P., and K. Fiedler. 2008. Global-scale modeling of groundwater recharge. *Hydrology and Earth System Sciences* 12:863–885.

Döll, P., K. Fiedler, and J. Zhang. 2009. Global-scale analysis of river flow alterations due to water withdrawals and reservoirs. *Hydrology and Earth System Sciences* 13:2413–2432.

Dombeck, M. 2003. The forgotten forest product: Water. *New York Times.* January 3, 2003.

Donohew, Z. 2009. Property rights and western United States water markets. *Australian Journal of Agricultural and Resource Economics* 53:85–103.

Doremus, H., and A. D. Tarlock. 2008. *Water War in the Klamath Basin: Macho Law, Combat Biology, and Dirty Politics.* Washington: Island Press.

dos Santos, M. A., L. P. Rosa, B. Sikar, E. Sikar, and E. O. dos Santos. 2006. Gross greenhouse gas fluxes from hydro-power reservoir compared to thermo-power plants. *Energy Policy* 34:481–488.

Doyle, M. W., E. H. Stanley, J. M. Harbor, and G. S. Grant. 2003. Dam removal in the United States: Emerging needs for science and policy. *Eos, Transactions, American Geophysical Union* 84:29, 32–33.

Duhigg, C. 2009a. Clean water laws are neglected, at a cost in suffering. *New York Times.* September 13, 2009.

Duhigg, C. 2009b. Cleansing the air at the expense of waterways. *New York Times.* October 13, 2009.

Duhigg, C. 2009c. Millions in U.S. drink dirty water, records show. *New York Times.* December 8, 2009.

Duhigg, C. 2009d. That tap water is legal but may be unhealthy. *New York Times.* December 17, 2009.

Dziegielewski, B., and T. Bik. 2006. *Water Use Benchmarks for Thermoelectric Power Generation.* Southern Illinois University Carbondale.

Eckstein, Y., and G. E. Eckstein. 2005. Transboundary aquifers: Conceptual models for development of international law. *Ground Water* 43:679–690.

Eisenberg, J. N. S., J. C. Scott, and T. Porco. 2007. Integrating disease control strategies: Balancing water sanitation and hygiene interventions to reduce diarrheal disease burden. *American Journal of Public Health* 97:846–852.

Emanuel, K. 2005. Increasing destructiveness of tropical cyclones over the past 30 years. *Nature* 436:686–688.

Environmental Protection Agency (EPA). 2001. *Technical Fact Sheet: Final Rule for Arsenic in Drinking Water.* EPA 815-F-00-016, available at www.epa.gov/safewater/arsenic/regulations_techfactsheet.html.

Environmental Protection Agency (EPA). 2002. *The Clean Water and Drinking Water Infrastructure Gap Analysis.* EPA-816-R-02-020.

Environmental Protection Agency (EPA). 2004. *Evaluation of Impacts to Underground Sources of Drinking Water by Hydraulic Fracturing of Coalbed Methane Reservoirs.* EPA 816-F-04-017, available at www.epa.gov/safewater/uic/pdfs/fs_cbmstudy_final_june2004.pdf.

Environmental Protection Agency (EPA). 2008. *2007 Annual Noncompliance Report National Pollutant Discharge Elimination System Non-Majors,* available at www.epa.gov/compliance/resources/reports/performance/cwa/cwa-npdes-non-majors-2007.pdf.

Environmental Protection Agency (EPA). 2009a. *2008 State Summary Data for Clean Water Act National Pollutant Discharge Elimination System Majors,* available at www.epa.gov/compliance/resources/reports/performance/cwa/cwa-npdes-majors-2008.pdf.

Environmental Protection Agency (EPA). 2009b. *2009 Edition of the Drinking Water Standards and Health Advisories.* EPA 822-R-09-011, available at www.epa.gov/waterscience/criteria/drinking/dwstandards2009.pdf.

Electric Power Research Institute (EPRI). 2002. *U.S. Water Consumption for Power Production: The Next Half Century.* Palo Alto, CA: EPRI.

Falkenmark, M. 1986. Fresh-water—Time for a modified approach. *Ambio* 15:192–200.

Falkenmark, M., and M. Lannerstad. 2005. Consumptive water use to feed humanity—Curing a blind spot. *Hydrology and Earth System Sciences* 9:15–28.

Falkenmark, M., and D. Molden. 2008. Wake up to realities of river basin closure. *Water Resources Development* 24:201–215.

Falkenmark, M., and J. Rockstrom. 2004. *Balancing Water for Humans and Nature: The New Approach in Ecohydrology.* London: Earthscan.

Fawell, J., K. Bailey, J. Chilton, E. Dahi, L. Fewtrell, and Y. Magara. 2006. *Fluoride in Drinking Water.* London: World Health Organization.

Fearnside, P. M. 2005. Brazil's Samuel Dam: Lessons for hydroelectric development policy and the environment in Amazonia. *Environmental Management* 35:1–19.

Feeley, T. J., T. J. Skone, G. J. Stlegel, A. McNemar, M. Nemeth, B. Schimmoller, J. T. Murph, and L. Manfredo. 2008. Water: A critical resource in the thermoelectric power industry. *Energy* 33:1–11.

Feitelson, E. 2000. The ebb and flow of Arab-Israeli water conflicts: Are past confrontations likely to resurface? *Water Policy* 2:343–363.

Few, R. 2007. Health and climatic hazards: Framing social research on vulnerability, response and adaptation. *Global Environmental Change—Human and Policy Dimensions* 17:281–295.

Fewtrell, L., R. B. Kaufmann, D. Kay, W. Enanoria, L. Haller, and J. M. Colford. 2005. Water, sanitation, and hygiene interventions to reduce diarrhoea in less developed countries: A systematic review and meta-analysis. *Lancet Infectious Diseases* 5:42–52.

Fischhendler, I. 2008a. Institutional conditions for IWRM: The Israeli case. *Ground Water* 46:91–102.

Fischhendler, I. 2008b. Ambiguity in transboundary environmental dispute resolution: The Israeli-Jordanian water agreement. *Journal of Peace Research* 45:91–109.

Fisher, F. M., A. Huber-Lee, I. Amir, S. Arlosoroff, Z. Eckstein, M. J. Haddadin, S. G. Hamati, A. M. Jarrar, A. F. Jayyousi, U. Shamir, and H. Wesseling. 2005. *Liquid Assets: An Economic Approach for Water Management and Conflict Resolution in the Middle East and Beyond.* Washington: Resources for the Future.

Food and Agriculture Organization (FAO). 2008. *The State of World Fisheries and Aquaculture 2008.* Rome: FAO.

Foster, S., and D. P. Loucks, eds. 2006. *Non-Renewable Groundwater Resources: A Guidebook on Socially-Sustainable Management for Water-Policy Makers.* Paris: United Nations Educational, Scientific, and Cultural Organization (UNESCO).

Freeze, R. A. 2000. *The Environmental Pendulum: A Quest for the Truth About Toxic Chemicals, Human Health, and Environmental Protection.* Berkeley: University of California Press.

Galloway, G. E. 2005. Corps of Engineers response to the changing national approach to floodplain management since the 1993 Midwest flood. *Journal of Contemporary Water Research and Education* 130:5–12.

Gaudin, S. 2006. Effect of price information on residential water demand. *Applied Economics* 38:383–393.

Ge, S. M., M. A. Liu, N. Lu, J. W. Godt, and G. Luo. 2009. Did the Zipingpu Reservoir trigger the 2008 Wenchuan earthquake? *Geophysical Research Letters* 36:DOI 10.1029/2009gl040349.

George, R. 2009. Yellow is the new green. *New York Times.* February 27, 2009.

Gerbens-Leenes, W., A. Y. Hoekstra, and T. H. van der Meer. 2009. The water footprint of bioenergy. *Proceedings of the National Academy of Sciences of the United States of America* 106:10219–10223.

Gerten, D., S. Rost, W. von Bloh, and W. Lucht. 2008. Causes of change in 20th century global river discharge. *Geophysical Research Letters* 35:5.

Gettleman, J. 2009. Even the camels are dying. *New York Times.* October 8, 2009.

Ghassemi, F., A. J. Jakeman, and H. A. Nix. 1995. *Salinisation of Land and Water Resources: Human Causes, Extent, Management, and Case Studies.* Wallingford, UK: CAB International.

Giordano, M. F., M. A. Giordano, and A. T. Wolf. 2005. International resource conflict and mitigation. *Journal of Peace Research* 42:47–65.

Glaister, D. 2004. Thirsty California starts to drink the Pacific. *The Guardian.* April 13, 2004.

Gleditsch, N. P., K. Furlong, H. Hegre, B. Lacina, and T. Owen. 2006. Conflicts over shared rivers: Resource scarcity or fuzzy boundaries? *Political Geography* 25:361–382.

Gleick, P. H. 2000. *The World's Water 2000–2001: The Biennial Report on Freshwater Resources.* Washington: Island Press.

Gleick, P. H. 2002. Water management—Soft water paths. *Nature* 418:373.

Gleick, P. H. 2003. Water use. *Annual Review of Environment and Resources* 28:275–314.

Gleick, P. H. 2006. Bottled water: An update. Pp. 169–174 in P. H. Gleick, ed. *The World's Water 2006–2007: The Biennial Report on Freshwater Resources.* Washington: Island Press.

Gleick, P. H., H. Cooley, and G. Wolff. 2006. With a grain of salt: An update on seawater desalination. Pp. 51–89 in P. H. Gleick, ed. *The World's Water 2006–2007: The Biennial Report on Freshwater Resources.* Washington: Island Press.

Gleick, P. H., D. Haasz, C. Henges-Jeck, V. Srinivasan, G. Wolff, K. K. Cushing, and A. Mann. 2003. *Waste Not, Want Not: The Potential for Urban Water Conservation in California.* Oakland, CA: Pacific Institute.

Gordon, L. J., W. Steffen, B. F. Jonsson, C. Folke, M. Falkenmark, and A. Johannessen. 2005. Human modification of global water vapor flows from the land surface. *Proceedings of the National Academy of Sciences of the United States of America* 102:7612–7617.

Gouldby, B., and P. Samuels. 2005. *Language of Risk—Project Definitions, FLOODsite Report T32-04-01,* available at www.floodsite.net/html/partner_area/project_docs/ FLOODsite_Language_of_Risk_v4_0_P1.pdf.

Govaerts, Y., and A. Lattanzio. 2008. Estimation of surface albedo increase during the eighties Sahel drought from Meteosat observations. *Global and Planetary Change* 64:139–145.

Green, S. T., M. J. Small, and E. A. Casman. 2009. Determinants of National Diarrheal Disease Burden. *Environmental Science & Technology* 43:993–999.

Griffin, R. C. 2006. *Water Resource Economics: The Analysis of Scarcity, Policies, and Projects.* Cambridge, MA: MIT Press.

Groisman, P. Y., R. W. Knight, D. R. Easterling, T. R. Karl, G. C. Hegerl, and V. A. N. Razuvaev. 2005. Trends in intense precipitation in the climate record. *Journal of Climate* 18:1326–1350.

Hall, D. G., K. S. Reeves, J. Brizzee, R. D. Lee, G. R. Carroll, and G. L. Sommers. 2006. *Feasibility Assessment of the Water Energy Resources of the United States for New Low Power and Small Hydro Classes of Hydroelectric Plants, DOE-ID-11263.* Idaho National Laboratory, available at www1.eere.energy.gov/windandhydro/pdfs/doe water-11263.pdf.

Halls, A. S., A. I. Payne, S. S. Alam, and S. K. Barman. 2008. Impacts of flood control schemes on inland fisheries in Bangladesh: Guidelines for mitigation. *Hydrobiologia* 609:45–58.

Hardin, G. 1968. The tragedy of the commons. *Science* 162:1243–1248.

Hashizume, M., Y. Wagatsuma, A. S. G. Faruque, T. Hayashi, P. R. Hunter, B. Armstrong, and D. A. Sack. 2008. Factors determining vulnerability to diarrhoea during and after severe floods in Bangladesh. *Journal of Water and Health* 6:323–332.

Hein, L., and N. de Ridder. 2006. Desertification in the Sahel: A reinterpretation. *Global Change Biology* 12:751–758.

Held, I. M., T. L. Delworth, J. Lu, K. L. Findell, and T. R. Knutson. 2005. Simulation of Sahel drought in the 20th and 21st centuries. *Proceedings of the National Academy of Sciences of the United States of America* 102:17891–17896.

Herrmann, S. M., A. Anyamba, and C. J. Tucker. 2005. Recent trends in vegetation dynamics in the African Sahel and their relationship to climate. *Global Environmental Change—Human and Policy Dimensions* 15:394–404.

Hoekstra, A. Y., and A. K. Chapagain. 2007. Water footprints of nations: Water use by people as a function of their consumption pattern. *Water Resources Management* 21:35–48.

Hoekstra, A. Y., and A. K. Chapagain. 2008. *Globalization of Water*. Oxford: Blackwell Publishing.

Hoff, H. 2009. Global water resources and their management. *Current Opinion in Environmental Sustainability* 1:141–147.

Holland, G. J., and P. J. Webster. 2007. Heightened tropical cyclone activity in the North Atlantic: Natural variability or climate trend? *Philosophical Transactions of the Royal Society—Mathematical, Physical and Engineering Sciences* 365:2695–2716.

Holland, S. P., and M. R. Moore. 2003. Cadillac Desert revisited: Property rights, public policy, and water-resource depletion. *Journal of Environmental Economics and Management* 46:131–155.

Hornberger, G. M., J. P. Raffensperger, P. L. Wiberg, and K. N. Eshleman. 1998. *Elements of Physical Hydrology*. Baltimore: The Johns Hopkins University Press. Available at www.npwrc.usgs.gov/resource/wetlands/wetloss/index.htm.

Hug, S. J., O. X. Leupin, and M. Berg. 2008. Bangladesh and Vietnam: Different groundwater compositions require different approaches to arsenic mitigation. *Environmental Science & Technology* 42:6318–6323.

Hulme, M. 2001. Climatic perspectives on Sahelian desiccation: 1973–1998. *Global Environmental Change—Human and Policy Dimensions* 11:19–29.

Human Rights Watch (HRW). 1995. *The Three Gorges Dam in China: Forced Resettlement, Suppression of Dissent and Labor Rights Concerns*. Available at www.hrw.org/legacy/reports/1995/China1.htm.

Hutson, S. S., N. L. Barber, J. F. Kenny, K. S. Linsey, D. S. M. Lumia, and M. A. Maupin. 2004. *Estimated Use of Water in the United States in 2000*. Reston, VA: United States Geological Survey.

Hutton, G., and J. Bartram. 2008. Global costs of attaining the Millennium Development Goal for water supply and sanitation. *Bulletin of the World Health Organization* 86:13–19.

Hutton, G., L. Haller, and J. Bartram. 2007. Global cost-benefit analysis of water supply and sanitation interventions. *Journal of Water and Health* 5:481–502.

Intergovernmental Panel on Climate Change (IPCC). 2007. *Climate Change 2007: Synthesis Report. Contribution of Working Groups I, II and III to the Fourth Assessment Report of the Intergovernmental Panel on Climate Change*. Geneva: IPCC, available at www.ipcc.ch/publications_and_data/ar4/syr/en/contents.html.

Izaguirre, A. K., and E. Pérard. 2009. *PPI Data Update 23*. Private Participation in Infrastructure Database.

Jensen, M. E. 2007. Beyond irrigation efficiency. *Irrigation Science* 25:233–245.

Jia, S. F., H. Yang, S. F. Zhang, L. Wang, and J. Xia. 2006. Industrial water use Kuznets curve: Evidence from industrialized countries and implications for developing countries. *Journal of Water Resources Planning and Management—ASCE* 132:183–191.

Jiménez, B., P. Drechsel, D. Koné, A. Bahri, L. Raschid-Sally, and M. Qadir. 2010. Wastewater, sludge and excreta use in developing countries: An overview. Pp. 3–28 in P. Drechsel, L. Raschid-Sally, M. Redwood, A. Bahri, and C. A. Scott, ed. *Wastewater Irrigation and Health: Assessing and Mitigating Risk in Low-Income Countries*. London: Earthscan.

Joint Monitoring Program (JMP). 2000. *Global Water Supply and Sanitation Assessment: 2000 Report*. New York and Geneva: UNICEF/WHO.

Joint Monitoring Program (JMP). 2008. *Progress on Drinking Water and Sanitation: Special Focus on Sanitation*. New York and Geneva: UNICEF/WHO.

Karr, J. R., and E. W. Chu. 1999. *Restoring Life in Running Waters: Better Biological Monitoring*. Washington: Island Press.

Keller, A., and J. Keller. 1995. *Effective efficiency: A water use efficiency concept for allocating freshwater resources, Discussion paper 22*. Arlington, VA: Center for Economic Policy Studies, Winrock International.

Kennard, M. J., A. H. Arthington, B. J. Pusey, and B. D. Harch. 2005. Are alien fish a reliable indicator of river health? *Freshwater Biology* 50:174–193.

Kenny, J. F., N. L. Barber, S. S. Hutson, K. S. Linsey, J. K. Lovelace, and M. A. Maupin. 2009. *Estimated Use of Water in the United States in 2005*. Reston, VA: United States Geological Survey.

Kitoh, A., A. Yatagai, and P. Alpert. 2008. First super-high-resolution model projection that the ancient "Fertile Crescent" will disappear this century. *Hydrological Research Letters* 2:1–4.

Knapp, A. K., C. Beier, D. D. Briske, A. T. Classen, Y. Luo, M. Reichstein, M. D. Smith, S. D. Smith, J. E. Bell, P. A. Fay, J. L. Heisler, S. W. Leavitt, R. Sherry, B. Smith, and E. Weng. 2008. Consequences of more extreme precipitation regimes for terrestrial ecosystems. *BioScience* 58:811–821.

Kumar, M. D., and O. P. Singh. 2005. Virtual water in global food and water policy making: Is there a need for rethinking? *Water Resources Management* 19:759–789.

Kunkel, K. E., D. R. Easterling, K. Redmond, and K. Hubbard. 2003. Temporal variations of extreme precipitation events in the United States: 1895–2000. *Geophysical Research Letters* 30:4.

Labat, D., Y. Godderis, J. L. Probst, and J. L. Guyot. 2004. Evidence for global runoff increase related to climate warming. *Advances in Water Resources* 27:631–642.

LaFraniere, S. 2009. Possible link between dam and China quake. *New York Times*. February 6, 2009.

Lambert, F. H., A. R. Stine, N. Y. Krakauer, and J. C. H. Chiang. 2008. How much will precipitation increase with global warming? *Eos, Transactions, American Geophysical Union* 89:193–200.

Lamichhane, K. M. 2007. On-site sanitation: A viable alternative to modern wastewater treatment plants. *Water Science and Technology* 55:433–440.

Landsea, C. W., B. A. Harper, K. Hoarau, and J. A. Knaff. 2006. Can we detect trends in extreme tropical cyclones? *Science* 313:452–454.

Langergraber, G., and E. Muellegger. 2005. Ecological sanitation: A way to solve global sanitation problems? *Environment International* 31:433–444.

Leopold, L. B. 1968. *Hydrology for Urban Land Planning—A Guidebook on the Hydrologic Effects of Urban Land Use, USGS Circular 554.* Reston, VA: United States Geological Survey.

Leopold, L. B., M. G. Wolman, and J. P. Miller. 1964. *Fluvial Processes in Geomorphology.* New York: Dover Publications, Inc.

Likens, G. E., and F. H. Bormann. 2008. *Biogeochemistry of a Forested Ecosystem,* 2nd ed. New York: Springer.

Lovins, A. B. 1977. *Soft Energy Paths: Toward a Durable Peace.* Cambridge, MA: Friends of the Earth / Ballinger Publishing.

MacKenzie, W. R., N. J. Hoxie, M. E. Proctor, M. S. Gradus, K. A. Blair, D. E. Peterson, J. J. Kazmierczak, D. G. Addiss, K. R. Fox, J. B. Rose, and J. P. Davis. 1994. A massive outbreak in Milwaukee of cryptosporidium infection transmitted through the public water supply. *New England Journal of Medicine* 331:161–167.

Mansur, E. T., and S. M. Olmstead. 2007. *The Value of Scarce Water: Measuring the Inefficiency of Municipal Regulations.* National Bureau of Economic Research (NBER) Working Paper #13513, Cambridge, MA: NBER.

Margat, J., S. Foster, and A. Droubi. 2006. Concept and importance of non-renewable resources. Pp. 13–24 in S. Foster and D. P. Loucks, eds. *Non-Renewable Groundwater Resources: A Guidebook on Socially-Sustainable Management for Water-Policy Makers.* Paris: United Nations Educational, Scientific, and Cultural Organization (UNESCO).

Mayer, P. W., W. B. DeOreo, E. M. Opitz, J. C. Kiefer, W. Y. Davis, B. Dziegielewski, and J. O. Nelson. 1999. *Residential End Uses of Water.* Denver: AWWA Research Foundation.

McCormack, M., G. J. Treloar, L. Palmowski, and R. Crawford. 2007. Modelling direct and indirect water requirements of construction. *Building Research & Information* 35:156–162.

McCully, P. 2001. *Silenced Rivers: The Ecology and Politics of Large Dams,* 2nd ed. London: Zed Books.

McGinnis, R. L., and M. Elimelech. 2008. Global challenges in energy and water supply: The promise of engineered osmosis. *Environmental Science & Technology* 42:8625–8629.

McKibben, B. 2007. *Deep Economy: The Wealth of Communities and the Durable Future.* New York: Henry Holt and Company.

Meko, D. M., C. A. Woodhouse, C. A. Baisan, T. Knight, J. J. Lukas, M. K. Hughes, and M. W. Salzer. 2007. Medieval drought in the upper Colorado River Basin. *Geophysical Research Letters* 34:5.

Merrett, S. 2003. Virtual water—A discussion—Virtual water and Occam's razor. *Water International* 28:103–105.

Merz, R., and G. Bloschl. 2008. Flood frequency hydrology: 1. Temporal, spatial, and causal expansion of information. *Water Resources Research* 44:DOI 10.1029/2007wr006744.

Milliman, J. D., K. L. Farnsworth, P. D. Jones, K. H. Xu, and L. C. Smith. 2008. Climatic and anthropogenic factors affecting river discharge to the global ocean, 1951–2000. *Global and Planetary Change* 62:187–194.

Milly, P. C. D., J. Betancourt, M. Falkenmark, R. M. Hirsch, Z. W. Kundzewicz, D. P. Lettenmaier, and R. J. Stouffer. 2008. Climate change—Stationarity is dead: Whither water management? *Science* 319:573–574.

Milly, P. C. D., K. A. Dunne, and A. V. Vecchia. 2005. Global pattern of trends in stream-flow and water availability in a changing climate. *Nature* 438:347–350.

Milly, P. C. D., R. T. Wetherald, K. A. Dunne, and T. L. Delworth. 2002. Increasing risk of great floods in a changing climate. *Nature* 415:514–517.

Mirza, M. M. Q. 2002. Global warming and changes in the probability of occurrence of floods in Bangladesh and implications. *Global Environmental Change—Human and Policy Dimensions* 12:127–138.

Morikawa, M., J. Morrison, and P. H. Gleick. 2009. Business reporting on water. Pp. 17–38 in P. H. Gleick, ed. *The World's Water 2008–2009: The Biennial Report on Freshwater Resources.* Washington: Island Press.

Mouawad, J., and C. Krauss. 2009. Dark side of a natural gas boom. *New York Times.* December 8, 2009.

Munich Re (2009). Natural catastrophes 2008: Analyses, assessments, positions, available at www.munichre.com/en/publications/default.aspx?id=1136.

Natan, T. E., and C. G. Miller. 1998. Are toxics release inventory reductions real? *Environmental Science & Technology* 32:368A–374A.

National Research Council (NRC). 2000. *Clean Coastal Waters: Understanding and Reducing the Effects of Nutrient Pollution.* Washington: National Academies Press.

National Research Council (NRC). 2003. *Oil in the Sea III.* Washington: National Academies Press.

National Research Council (NRC). 2007. *Colorado River Basin Water Management: Evaluating and Adjusting to Hydroclimatic Variability.* Washington: National Academies Press.

National Research Council (NRC). 2008a. *Desalination: A National Perspective.* Washington: National Academies Press.

National Research Council (NRC). 2008b. *Water Implications of Biofuels Production in the United States.* Washington: National Academies Press.

Nilsson, C., C. A. Reidy, M. Dynesius, and C. Revenga. 2005. Fragmentation and flow regulation of the world's large river systems. *Science* 308:405–408.

Nohara, D., A. Kitoh, M. Hosaka, and T. Oki. 2006. Impact of climate change on river discharge projected by multimodel ensemble. *Journal of Hydrometeorology* 7:1076–1089.

Olmstead, S. M., and R. N. Stavins. 2009. Comparing price and nonprice approaches to urban water conservation. *Water Resources Research* 45:10.

Omernick, J. M., and R. G. Bailey. 1997. Distinguishing between watersheds and ecoregions. *Journal of the American Water Resources Association* 33:935–949.

Organisation for Economic Co-operation and Development (OECD). 2009. *Managing Water for All: An OECD Perspective on Pricing and Financing.* Paris: OECD.

Ozsvath, D. L. 2009. Fluoride and environmental health: A review. *Reviews in Environmental Science and Bio/Technology* 8:59–79.

Palaniappan, M., P. H. Gleick, C. Hunt, and V. Srinivasan. 2004. Water privatization principles and practices. Pp. 45–78 in P. H. Gleick, ed. *The World's Water 2004–2005: The Biennial Report on Freshwater Resources.* Washington: Island Press.

Peng, S. B., J. L. Huang, J. E. Sheehy, R. C. Laza, R. M. Visperas, X. H. Zhong, G. S. Centeno, G. S. Khush, and K. G. Cassman. 2004. Rice yields decline with higher night temperature from global warming. *Proceedings of the National Academy of Sciences of the United States of America* 101:9971–9975.

Pérard, E. 2009. Water supply: Public or private? An approach based on cost of funds, transaction costs, efficiency and political costs. *Policy and Society* 27:193–219.

Petit, C., T. Scudder, and E. Lambin. 2001. Quantifying processes of land-cover change by remote sensing: Resettlement and rapid land-cover changes in south-eastern Zambia. *International Journal of Remote Sensing* 22:3435–3456.

Piao, S. L., P. Friedlingstein, P. Ciais, N. de Noblet-Ducoudre, D. Labat, and S. Zaehle. 2007. Changes in climate and land use have a larger direct impact than rising $CO_2$ on global river runoff trends. *Proceedings of the National Academy of Sciences of the United States of America* 104:15242–15247.

Pielke, R. A., Jr., M. W. Downton, and J. Z. Barnard Miller. 2002. *Flood Damage in the United States, 1926–2000: A Reanalysis of National Weather Service Estimates.* Boulder, CO: UCAR.

Pielke, R. A., C. Landsea, M. Mayfield, J. Laver, and R. Pasch. 2005. Hurricanes and global warming. *Bulletin of the American Meteorological Society* 86:1571-+.

Plater, Z. J. B. 2004. Endangered Species Act lessons over 30 years and the legacy of the snail darter, a small fish in a porkbarrel. *Environmental Law* 34:289–308.

Poff, N. L., J. D. Allan, M. B. Bain, J. R. Karr, K. L. Prestegaard, B. D. Richter, R. E. Sparks, and J. C. Stromberg. 1997. The natural flow regime: A paradigm for river conservation and restoration. *BioScience* 47:769–784.

Poff, N. L., J. D. Olden, D. M. Merritt, and D. M. Pepin. 2007. Homogenization of regional river dynamics by dams and global biodiversity implications. *Proceedings of the National Academy of Sciences of the United States of America* 104:5732–5737.

Poff, N. L., B. D. Richter, A. H. Arthington, S. E. Bunn, R. J. Naiman, E. Kendy, M. Acreman, C. Apse, B. P. Bledsoe, M. C. Freeman, J. Henriksen, R. B. Jacobson, J. G. Kennen, D. M. Merritt, J. H. O'Keeffe, J. D. Olden, K. Rogers, R. E. Tharme, and A. Warner. 2010. The ecological limits of hydrologic alteration (ELOHA): A new framework for developing regional environmental flow standards. *Freshwater Biology* 55:147–170.

Postel, S. 1999. *Pillar of Sand: Can the Irrigation Miracle Last?* New York: W. W. Norton & Company.

Potere, D., and A. Schneider. 2007. A critical look at representations of urban areas in global maps. *GeoJournal* 69:55–80.

Prüss, A., D. Kay, L. Fewtrell, and J. Bartram. 2002. Estimating the burden of disease from water, sanitation, and hygiene at a global level. *Environmental Health Perspectives* 110:537–542.

Prüss-Üstün, A., R. Bos, F. Gore, and J. Bartram. 2008. *Safer Water, Better Health: Costs, Benefits and Sustainability of Interventions to Protect and Promote Health.* Geneva: World Health Organization.

Redman, R. L., C. A. Nenn, D. Eastwood, and M. H. Gorelick. 2007. Pediatric emergency department visits for diarrheal illness increased after release of undertreated sewage. *Pediatrics* 120:E1472–E1475.

Renzetti, S. 1999. Municipal water supply and sewage treatment: Costs, prices, and distortions. *Canadian Journal of Economics—Revue Canadienne D Economique* 32:688–704.

Revenga, C., J. Brunner, N. Henninger, K. Kassem, and R. Payne. 2000. *Pilot Analysis of Global Ecosystems: Freshwater Systems.* Washington: World Resources Institute.

Reynolds, K. A., K. D. Mena, and C. P. Gerba. 2008. Risk of waterborne illness via drinking water in the United States. *Reviews of Environmental Contamination and Toxicology* 192:117–158.

Richter, B. D., J.V. Baumgartner, J. Powell, and D. P. Braun. 1996. A method for assessing hydrologic alteration within ecosystems. *Conservation Biology* 10:1163–1174.

Rock, M. T. 1998. Freshwater use, freshwater scarcity, and socioeconomic development. *Journal of Environment and Development* 7:278.

Rosenthal, E. 2009. In Bolivia, water and ice tell of climate change. *New York Times.* December 15, 2009.

Rosgen, D. 1996. *Applied River Morphology.* Pagosa Springs, CO: Wildland Hydrology.

Rost, S., D. Gerten, A. Bondeau, W. Lucht, J. Rohwer, and S. Schaphoff. 2008. Agricultural green and blue water consumption and its influence on the global water system. *Water Resources Research* 44:10.1029/2007wr006331.

Roth, D., and J. Warner. 2008. Virtual water: Virtuous impact? The unsteady state of virtual water. *Agriculture and Human Values* 25:257–270.

Sachs, J. D. 2009. *Common Wealth: Economics for a Crowded Planet.* New York: Penguin Press.

Samuels, P., F. Klijn, and J. Dijkman. 2006. An analysis of the current practice of policies on river flood risk management in different countries. *Irrigation and Drainage* 55:S141–S150.

Scarborough, B., and H. L. Lund. 2007. *Saving Our Streams: Harnessing Water Markets.* Bozeman, MT: Property and Environment Research Center.

Schmidt, W. P., and S. Cairncross. 2009. Household water treatment in poor populations: Is there enough evidence for scaling up now? *Environmental Science & Technology* 43:986–992.

Scudder, T. 2005. *The Future of Large Dams: Dealing With Social, Environmental, Institutional and Political Costs.* London: Earthscan.

Seager, R., M. F. Ting, I. Held, Y. Kushnir, J. Lu, G. Vecchi, H. P. Huang, N. Harnik, A. Leetmaa, N. C. Lau, C. H. Li, J. Velez, and N. Naik. 2007. Model projections of an imminent transition to a more arid climate in southwestern North America. *Science* 316:1181–1184.

Seckler, D., U. Armarasinghe, D. Molden, R. de Silva, and R. Barker. 1998. *World Water Demand and Supply, 1990 to 2025: Scenarios and Issues.* Colombo: International Water Management Institute.

Sehlke, G. 2009. What is "the energy-water nexus"? *Journal of Contemporary Water Research and Education* 143:1–2.

Serageldin, I. 2009. Water: Conflicts set to arise within as well as between states. *Nature* 459:163.

Shah, T., J. Burke, and K. Billholth. 2007. Groundwater: A global assessment of scale and significance. Pp. 395–423 in Comprehensive Assessment of Water Management in Agriculture, ed. *Water for Food, Water for Life: A Comprehensive Assessment of Water Management in Agriculture.* London and Colombo: Earthscan and International Water Management Institute.

Shao, X., H. Wang, and Z. Wang. 2003. Interbasin transfer projects and their implications: A China case study. *International Journal of River Basin Management* 1:5–14.

Shiklomanov, I. A. 1999. World water resources and their use, a joint SHI/UNESCO product, webworld.unesco.org/water/ihp/db/shiklomanov/.

Shiklomanov, I. A., and J. C. Rodda. 2003. *World Water Resources at the Beginning of the 21st Century*. Cambridge: Cambridge University Press.

Smakhtin, V., C. Revenga, and P. Döll. 2004. A pilot global assessment of environmental water requirements and scarcity. *Water International* 29:307–317.

Small, C., and R. J. Nicholls. 2003. A global analysis of human settlement in coastal zones. *Journal of Coastal Research* 19:584–599.

Smith, A. H., E. O. Lingas, and M. Rahman. 2000. Contamination of drinking-water by arsenic in Bangladesh: A public health emergency. *Bulletin of the World Health Organization* 78:1093–1103.

Solley, W. B., R. R. Pierce, and H. A. Perlman. 1998. *Estimated Use of Water in the United States in 1995, Circular 1200*. Reston, VA: United States Geological Survey.

Speth, J. G. 2008. *The Bridge at the Edge of the World: Capitalism, the Environment, and Crossing from Crisis to Sustainability*. New Haven, CT: Yale University Press.

St. Louis, V. L., C. A. Kelly, E. Duchemin, J. W. M. Rudd, and D. M. Rosenberg. 2000. Reservoir surfaces as sources of greenhouse gases to the atmosphere: A global estimate. *BioScience* 50:766–775.

Stein, B. A., and S. R. Flack. 1997. *1997 Species Report Card: The State of U.S. Plants and Animals*. Arlington, VA: The Nature Conservancy.

Stevens, W. K. 2000. Megadrought appears to loom in Africa. *New York Times*. February 8, 2000.

Swindell, G. S. 2007. *Powder River Basin Coalbed Methane Wells—Reserves and Rates*. SPE Rocky Mountain Oil & Gas Technology Symposium. Denver, CO, Society of Petroleum Engineers.

Tharme, R. E. 2003. A global perspective on environmental flow assessment: Emerging trends in the development and application of environmental flow methodologies for rivers. *River Research and Applications* 19:397–441.

Thatte, C. D. 2007. Inter-basin water transfer (IBWT) for the augmentation of water resources in India: A review of needs, plans, status and prospects. *International Journal of Water Resources Development* 23:709–725.

Thomas, D. H. L., and W. M. Adams. 1999. Adapting to dams: Agrarian change downstream of the Tiga Dam, Northern Nigeria. *World Development* 27:919–935.

Thompson, J. R., and G. Polet. 2000. Hydrology and land use in a Sahelian floodplain wetland. *Wetlands* 20:639–659.

Thompson, S. A. 1999. *Water Use, Management, and Planning in the United States*. San Diego: Academic Press.

Tiwari, V. M., J. Wahr, and S. Swenson. 2009. Dwindling groundwater resources in northern India, from satellite gravity observations. *Geophysical Research Letters* 36:10.1029/2009gl039401.

Traister, E., and S. C. Anisfeld. 2006. Variability of indicator bacteria at different time scales in the upper Hoosic River watershed. *Environmental Science & Technology* 40:4990–4995.

Trenberth, K. E., A. G. Dai, R. M. Rasmussen, and D. B. Parsons. 2003. The changing character of precipitation. *Bulletin of the American Meteorological Society* 84:1205–1217.

Trout Unlimited. 2006. *A Glass Half Full: The Future of Water in New England*. Trout

Unlimited, available at www.tu.org/atf/cf/%7B0D18ECB7-7347-445B-A38E -65B282BBBD8A%7D/A%20GLASS%20HALF%20FULL.pdf.

Turner, R. E. 1997. Wetland loss in the northern Gulf of Mexico: Multiple working hypotheses. *Estuaries* 20:1–13.

Uchida, H., and K. Ando. 2007. Adaptive agricultural system to dynamic water condition in a low-lying area of Bangladesh. *Jarq-Japan Agricultural Research Quarterly* 41:25–30.

United Nations Development Programme (UNDP). 2006. *Human Development Report 2006. Beyond Scarcity: Power, Poverty, and the Global Water Crisis.* New York: UNDP.

United Nations Environment Programme (UNEP). 2002. *Vital Water Graphics,* available at www.unep.org/vitalwater.

United States Climate Change Science Program (USCCSP). 2008. *Weather and Climate Extremes in a Changing Climate.*

United States Department of Energy (DOE). 2006. *Energy Demands on Water Resources: Report to Congress on the Interdependency of Energy and Water.*

United States Geological Survey (USGS). 2009. National Water Information System, http://waterdata.usgs.gov/nwis.

Vannote, R. L., G. W. Minshall, K. W. Cummins, J. R. Sedell, and C. E. Cushing. 1980. The River Continuum Concept. *Canadian Journal of Fisheries and Aquatic Sciences* 37:130–137.

Venkatachalam, A. J., A. R. G. Price, S. Chandrasekara, and S. S. Sellamuttu. 2009. Risk factors in relation to human deaths and other tsunami (2004) impacts in Sri Lanka: The fishers'-eye view. *Aquatic Conservation—Marine and Freshwater Ecosystems* 19:57–66.

Verschuren, D., K. R. Laird, and B. F. Cumming. 2000. Rainfall and drought in equatorial east Africa during the past 1,100 years. *Nature* 403:410–414.

Vickers, A. 2001. *Handbook of Water Use and Conservation.* Amherst, MA: Waterplow Press.

Voith, M. 2008. The other scarce resource. *Chemical and Engineering News.* October 6, 2008:12–19.

Vorosmarty, C. J., E. M. Douglas, P. A. Green, and C. Revenga. 2005. Geospatial indicators of emerging water stress: An application to Africa. *Ambio* 34:230–236.

Vorosmarty, C. J., P. Green, J. Salisbury, and R. B. Lammers. 2000. Global water resources: Vulnerability from climate change and population growth. *Science* 289:284–288.

Vorosmarty, C. J., and D. Sahagian. 2000. Anthropogenic disturbance of the terrestrial water cycle. *BioScience* 50:753–765.

Vorosmarty, C. J., K. P. Sharma, B. M. Fekete, A. H. Copeland, J. Holden, J. Marble, and J. A. Lough. 1997. The storage and aging of continental runoff in large reservoir systems of the world. *Ambio* 26:210–219.

Vrba, J., and J. van der Gun. 2004. *The World's Groundwater Resources: Contribution to Chapter 4 of WWDR-2.* Utrecht: International Groundwater Resources Assessment Centre.

Wahl, R. W. 1989. *Markets for Federal Water: Subsidies, Property Rights, and the Bureau of Reclamation.* Washington: RFF Press.

Walsh, C. J., A. H. Roy, J. W. Feminella, P. D. Cottingham, P. M. Groffman, and R. P. Morgan. 2005. The urban stream syndrome: Current knowledge and the search for a cure. *Journal of the North American Benthological Society* 24:706–723.

Ward, F. A., and M. Pulido-Velazquez. 2008. Water conservation in irrigation can increase water use. *Proceedings of the National Academy of Sciences of the United States of America* 105:18215–18220.

Webber, M. E. 2008. Catch-22: Water vs. Energy. *Scientific American*. October 2008.

Weiskel, P. K., R. M. Vogel, P. A. Steeves, P. J. Zarriello, L. A. DeSimone, and K. G. Ries. 2007. Water use regimes: Characterizing direct human interaction with hydrologic systems. *Water Resources Research* 43:11.

Wentz, F. J., L. Ricciardulli, K. Hilburn, and C. Mears. 2007. How much more rain will global warming bring? *Science* 317:233–235.

Wheater, H. S. 2006. Flood hazard and management: A UK perspective. *Philosophical Transactions of the Royal Society A—Mathematical Physical and Engineering Sciences* 364:2135–2145.

Wildman, L. A. S., and J. G. MacBroom. 2005. The evolution of gravel bed channels after dam removal: Case study of the Anaconda and Union City Dam removals. *Geomorphology* 71:245–262.

Willmott, C. J., and J. J. Feddema. 1992. A more rational climatic moisture index. *Professional Geographer* 44:84–88.

Wittfogel, K. A. 1957. *Oriental Despotism: A Comparative Study of Total Power*. New Haven, CT: Yale University Press.

Wolf, A. T. 2002. *Atlas of International Freshwater Agreements*. Nairobi: United Nations Environment Programme (UNEP).

Wolf, A. T. 2007. Shared waters: Conflict and cooperation. *Annual Review of Environment and Resources* 32:241–269.

Wolf, A. T., J. A. Natharius, J. J. Danielson, B. S. Ward, and J. K. Pender. 1999. International river basins of the world. *International Journal of Water Resources Development* 15:387–427.

Wolfe, J. R., R. A. Goldstein, J. S. Maulbetsch, and C. R. McGowin. 2009. An electric power industry perspective on water use efficiency. *Journal of Contemporary Water Research and Education* 143:30–34.

Wolff, G., and P. H. Gleick. 2002. The soft path for water. Pp. 1–32 in P. H. Gleick, ed. *The World's Water 2002–2003: The Biennial Report on Freshwater Resources*. Washington: Island Press.

World Commission on Dams (WCD). 2000. *Dams and Development: A New Framework for Decision-Making*. London: Earthscan.

World Health Organization (WHO) and United Nations Children's Fund (UNICEF). 2006. Protecting and promoting human health. Pp. 203–240 in *Water: A Shared Responsibility: The United Nations World Water Development Report 2*. Paris: UNESCO, and New York: Berghahn Books.

World Water Assessment Programme (WWAP). 2009. *The United Nations World Water Development Report 3: Water in a Changing World*. Paris: UNESCO.

Yang, H., L. Wang, K. C. Abbaspour, and A. J. B. Zehnder. 2006. Virtual water trade: An assessment of water use efficiency in the international food trade. *Hydrology and Earth System Sciences* 10:443–454.

Yang, H., P. Reichert, K. C. Abbaspour, and A. J. B. Zehnder. 2003. A water resources threshold and its implications for food security. *Environmental Science & Technology* 37:3048–3054.

Yoffe, S., G. Fiske, M. Giordano, K. Larson, K. Stahl, and A. T. Wolf. 2004. Geography of international water conflict and cooperation: Data sets and applications. *Water Resources Research* 40:12.

Yoffe, S., A. T. Wolf, and M. Giordano. 2003. Conflict and cooperation over international freshwater resources: Indicators of basins at risk. *Journal of the American Water Resources Association* 39:1109–1126.

Zeitoun, M., C. Messerschmid, and S. Attili. 2009. Asymmetric abstraction and allocation: The Israeli–Palestinian water pumping record. *Ground Water* 47:146–160.

Zeitoun, M., and N. Mirumachi. 2008. Transboundary water interaction I: Reconsidering conflict and cooperation. *International Environmental Agreements-Politics, Law and Economics* 8:297–316.

Zhou, Y., and R. S. J. Tol. 2005. Evaluating the costs of desalination and water transport. *Water Resources Research* 41:10.

# Recommended Readings

I provide three kinds of recommended readings:

- Basic textbooks covering different aspects of water resources. This list, organized by subject, will be useful to those who want more background in the different academic disciplines necessary for understanding water issues.
- Ten important water books of the last decade. This list, organized alphabetically by author, is my subjective judgment of books that introduced important new ideas or provided a valuable synthesis.
- Chapter readings. For each chapter, I have selected a small number of important articles or reports that go into greater depth on some aspect of the chapter. These are, generally speaking, the readings that I use in my Water Resource Management class at Yale and that I recommend for use with this book. They are organized by the order in which subjects appear in the chapter.

*Basic Texts on Aspects of Water Resources*

hydrology: Hornberger, G. M., J. P. Raffensperger, P. L. Wiberg, and K. N. Eshleman. 1998. *Elements of Physical Hydrology*. Baltimore: The Johns Hopkins University Press.

water quality: Laws, E. A. 2000. *Aquatic Pollution: An Introductory Text*. New York: John Wiley & Sons.

river ecology: Allan, J. D. 1995. *Stream Ecology: Structure and Function of Running Waters*. Dordrecht: Kluwer Academic Publishers.

river science and management: MacBroom, J. G. 1998. *The River Book*. Hartford: Connecticut Department of Environmental Protection.

fluvial geomorphology: Charlton, R. 2008. *Fundamentals of Fluvial Geomorphology*. New York: Routledge; and

Rosgen, D. 1996. *Applied River Morphology*. Pagosa Springs, CO: Wildland Hydrology.

environmental economics: Keohane, N. O., and S. M. Olmstead. 2007. *Markets and the Environment*. Washington: Island Press.

water economics: Griffin, R. C. 2006. *Water Resource Economics: The Analysis of Scarcity, Policies, and Projects*. Cambridge, MA: MIT Press; and

Fisher, F. M., A. Huber-Lee, I. Amir, S. Arlosoroff, Z. Eckstein, M. J. Haddadin, S. G. Hamati, A. M. Jarrar, A. F. Jayyousi, U. Shamir, and H. Wesseling. 2005. *Liquid Assets: An Economic Approach for Water Management and Conflict Resolution in the Middle East and Beyond*. Washington: Resources for the Future.

environmental law: *Environmental Law Handbook*. Rockville, MD: Government Institutes (no section on Endangered Species Act); and

*Environmental Regulation: Law, Science, and Policy*. New York: Aspen Law and Business.

*Ten Important Water Books of the Last Decade*

Allan, J. A. 2001. *The Middle East Water Question: Hydropolitics and the Global Economy.* London: I B Tauris and Co.

Barlow, M. 2007. *Blue Covenant: The Global Water Crisis and the Coming Battle for the Right to Water.* New York: The New Press.

Falkenmark, M., and J. Rockstrom. 2004. *Balancing Water for Humans and Nature: The New Approach in Ecohydrology.* London: Earthscan.

Gleick, P. H. 2009. *The World's Water 2008–2009: The Biennial Report on Freshwater Resources.* Washington: Island Press (and previous books in the series).

Glennon, R. 2009. *Unquenchable: America's Water Crisis and What to Do About It.* Washington: Island Press.

Lenton, R., and M. Muller, eds. 2009. *Integrated Water Resources Management in Practice: Better Water Management for Development.* London: Earthscan.

McCully, P. 2001. *Silenced Rivers: The Ecology and Politics of Large Dams,* 2nd ed. London: Zed Books.

Pearce, F. 2006. *When the Rivers Run Dry: Water—The Defining Crisis of the Twenty-First Century.* Boston: Beacon Press.

Postel, S., and B. Richter. 2003. *Rivers for Life: Managing Water for People and Nature.* Washington: Island Press.

Scudder, T. 2005. *The Future of Large Dams: Dealing with Social, Environmental, Institutional and Political Costs.* London: Earthscan.

*Selected Chapter Readings*

CHAPTER 1

Gleick, P. H. 2003. Global freshwater resources: Soft-path solutions for the 21st century. *Science* 302:1524–1528.

CHAPTER 2

Omernick, J. M., and R. G. Bailey. 1997. Distinguishing between watersheds and ecoregions. *Journal of the American Water Resources Association* 33:935–949.

CHAPTER 3

Gleick, P. H. 2003. Water use. *Annual Review of Environment and Resources* 28:275–314.

Hoekstra, A. Y., and A. K. Chapagain. 2007. Water footprints of nations: Water use by people as a function of their consumption pattern. *Water Resources Management* 21:35–48.

Kenny, J. F., N. L. Barber, S. S. Hutson, K. S. Linsey, J. K. Lovelace, and M. A. Maupin. 2009. *Estimated Use of Water in the United States in 2005.* Reston, VA: United States Geological Survey.

CHAPTER 4

Brouwer, R., S. Akter, L. Brander, and E. Haque. 2007. Socioeconomic vulnerability and adaptation to environmental risk: A case study of climate change and flooding in Bangladesh. *Risk Analysis* 27:313–326.

Federal Emergency Management Agency (FEMA). 2002. National Flood Insurance Program—Program Description, available at www.fema.gov/library/viewRecord. do?id=1480.

McPhee, J. 1989. *The Control of Nature*. New York: Farrar, Straus and Giroux; Chapter 1 (Atchafalaya).

CHAPTER 5

Rijsberman, F. R. 2006. Water scarcity: Fact or fiction? *Agricultural Water Management* 80:5–22.

Falkenmark, M., and D. Molden. 2008. Wake up to realities of river basin closure. *Water Resources Development* 201–215.

CHAPTER 6

Milly, P. C. D., J. Betancourt, M. Falkenmark, R. M. Hirsch, Z. W. Kundzewicz, D. P. Lettenmaier, and R. J. Stouffer. 2008. Climate change—Stationarity is dead: Whither water management? *Science* 319:573–574.

Kundzewicz, Z. W., L. J. Mata, N. W. Arnell, P. Doll, B. Jimenez, K. Miller, T. Oki, Z. Sen, and I. Shiklomanov. 2008. The implications of projected climate change for freshwater resources and their management. *Hydrological Sciences Journal—Journal Des Sciences Hydrologiques* 53:3–10.

Alcamo, J., M. Florke, and M. Marker. 2007. Future long-term changes in global water resources driven by socio-economic and climatic changes. *Hydrological Sciences Journal—Journal Des Sciences Hydrologiques* 52:247–275.

Bruijnizeel, L. A. 2004. Hydrological functions of tropical forests: Not seeing the soil for the trees? *Agriculture, Ecosystems and Environment* 104:185–228.

CHAPTER 7

World Commission on Dams (WCD). 2000. *Dams and Development: A New Framework for Decision-Making*. London: Earthscan, executive summary, available at www.unep.org/dams/WCD/report.asp.

Yang, H., L. Wang, K. C. Abbaspour, and A. J. B. Zehnder. 2006. Virtual water trade: An assessment of water use efficiency in the international food trade. *Hydrology and Earth System Sciences* 10:443–454.

National Research Council (NRC). 2008. *Desalination: A National Perspective*. Washington: National Academies Press, summary, available at http://books.nap.edu/openbook.php?record_id=12184&page=1.

Brown, C., and U. Lall. 2006. Water and economic development: The role of variability and a framework for resilience. *Natural Resources Forum* 30:306–317.

CHAPTER 8

Poff, N. L., B. D. Richter, A. H. Arthington, S. E. Bunn, R. J. Naiman, E. Kendy, M. Acreman, C. Apse, B. P. Bledsoe, M. C. Freeman, J. Henriksen, R. B. Jacobson, J. G. Kennen, D. M. Merritt, J. H. O'Keeffe, J. D. Olden, K. Rogers, R. E. Tharme, and A. Warner. 2010. The ecological limits of hydrologic alteration (ELOHA): A new framework for developing regional environmental flow standards. *Freshwater Biology* 55:147–170.

Döll, P., K. Fiedler, and J. Zhang. 2009. Global-scale analysis of river flow alterations due to water withdrawals and reservoirs. *Hydrology and Earth System Sciences* 13:2413–2432.

Walsh, C. J., A. H. Roy, J. W. Feminella, P. D. Cottingham, P. M. Groffman, and R. P. Morgan. 2005. The urban stream syndrome: Current knowledge and the search for a cure. *Journal of the North American Benthological Society* 24:706–723.

Duhigg, C. 2009. As sewers fill, waste poisons waterways. *New York Times.* November 23, 2009.

CHAPTER 9

Duhigg, C. 2009. That tap water is legal but may be unhealthy. *New York Times.* December 17, 2009.

Joint Monitoring Program (JMP). 2008. *Progress on Drinking Water and Sanitation: Special Focus on Sanitation.* New York and Geneva: UNICEF / WHO.

Schmidt, W. P., and S. Cairncross. 2009. Household water treatment in poor populations: Is there enough evidence for scaling up now? *Environmental Science & Technology* 43:986–992.

Cooley, H., T. Hutchins-Cabibi, M. Cohen, P. H. Gleick, and M. Heberger. 2007. *Hidden Oasis: Water Conservation and Efficiency in Las Vegas.* Oakland, CA: Pacific Institute.

CHAPTER 10

Comprehensive Assessment of Water Management in Agriculture (CAWMA). 2007. *Water for Food, Water for Life: A Comprehensive Assessment of Water Management in Agriculture.* London and Colombo: Earthscan and International Water Management Institute, summary for decision makers, available from www.iwmi.cgiar.org/Assessment/.

Jensen, M. E. 2007. Beyond irrigation efficiency. *Irrigation Science* 25:233–245.

Ward, F. A., and M. Pulido-Velazquez. 2008. Water conservation in irrigation can increase water use. *Proceedings of the National Academy of Sciences of the United States of America* 105:18215–18220.

Duhigg, C. 2009. Health ills abound as farm runoff fouls wells. *New York Times.* September 18, 2009.

CHAPTER 11

Webber, M. E. 2008. Catch-22: Water vs. energy. *Scientific American.* October.

National Research Council (NRC). 2008. *Water Implications of Biofuels Production in the United States.* Washington: National Academies Press.

CHAPTER 12

Dellapenna, J. W. 2005. Markets for water: Time to put the myth to rest? *Journal of Contemporary Water Research and Education* 131:33–41.

Glennon, R. 2005. Water scarcity, marketing, and privatization. *Texas Law Review* 83:1873–1902.

Hardin, G. 1968. Tragedy of commons. *Science* 162:1243–1248.

CHAPTER 13

Wolf, A. T. 2007. Shared waters: Conflict and cooperation. *Annual Review of Environment and Resources* 32:241–269.

Fischhendler, I. 2008b. Ambiguity in transboundary environmental dispute resolution: The Israeli-Jordanian water agreement. *Journal of Peace Research* 45:91–109.

Ruhl, J. B. 2003. Equitable apportionment of ecosystem services: New water law for a new water age. *Journal of Land Use & Environmental Law* 19:47.

# Index